Maximizing Amazon Astro

Harnessing AI Robotics for Safer and Smarter Living

Markos Fletcher

Copyright

All rights reserved. No part of this book may be reproduced or used in any manner without the prior written permission of the copyright owner, except for the use of brief quotations in a book review.

While the advice and information in this book are believed to be true and accurate at the date of publication, neither the authors nor the editors nor the publisher can accept any legal responsibility for any errors or omissions that may be made. The publisher makes no warranty, express or implied, with respect to the material contained herein.

Printed on acid-free paper.

©2024

Contents

- Copyright 1
- Introduction 4
- What is Amazon Astro? 4
- Key Features and Capabilities of Amazon Astro 6
- Unboxing Your Amazon Astro 9
- Setting Up Astro 10
- Astro Application 14
- The Design and Hardware Breakdown 30
- How Astro Connects with Current Smart Home Devices and Alexa 41
- Astro in Action 52
- Astro Fun & Games 60
- How Astro can entertain and educate kids through activities 68
- The Function of Astro in Daily Life 76
- Maintaining and Customizing Astro 84
- Taking Control of Your Data and Privacy 93
- Troubleshooting and Common Issues 103
- How to Factory Reset Amazon Astro 117
- Keeping Astro Up to Date 117
- The Future of Astro 126
- Living with a Robot Companion 133

How Astro Works with Other Smart Home Appliances 142

Astro Safety and Compliance Information 150

HOW TO USE AND CARE FOR YOUR DEVICE 155

ABOUT THE AUTHOR 160

Introduction

In the ever-changing world of technology, Amazon has consistently been at the forefront of innovation, offering revolutionary solutions to everyday problems. Amazon Astro, the groundbreaking household robot designed to change the way we interact with our living spaces, is one of their most recent pioneering innovations. Astro, with its superior robotics, artificial intelligence, and seamless integration with the Amazon ecosystem, is poised to become an integral part of today's smart home. This introduction goes into Amazon Astro's capabilities, outlining its features and functionalities including the disruptive potential it has for home automation and personal assistance.

What is Amazon Astro?

Amazon Astro is an adaptive, autonomous household robot designed to help with a variety of tasks around the house. This small but powerful device is packed with a variety of sensors, cameras, and a display screen, allowing it to explore your home, communicate with people, and perform a wide range of functions. Astro is intended to give convenience, security, and entertainment, making it a complete solution for modern living.

Amazon Astro represents a significant leap forward in the realm of home automation and robotics. Its ability to integrate seamlessly with the Amazon ecosystem and other smart home devices makes it a central hub for managing and enhancing daily life. As technology continues to advance, Astro is expected to evolve, offering even more sophisticated features and capabilities.

Amazon Astro is more than just a gadget; it is a glimpse into the future of smart living. Its diverse range of functionalities, from home security to personal assistance, positions it as a versatile and invaluable addition to any home. As you explore the capabilities of Amazon Astro, you will discover how it can transform your living space into a more secure, convenient, and enjoyable environment.

Key Features and Capabilities of Amazon Astro

1. **Autonomous Navigation and Mapping:** Astro navigates your home automatically, using powerful sensors and mapping technology. It creates a precise map of your living area, allowing it to navigate effectively and avoid obstructions. Whether you live in a small apartment or a huge house, Astro adapts to your surroundings, guaranteeing smooth and safe movement.

2. **Home Monitoring and Security:** One of the most notable benefits of Amazon Astro is its capacity to improve home security. Astro, equipped with high-definition cameras, can patrol your home and send real-time video feeds to your smartphone. It detects odd activity and sends notifications, providing peace of mind whether you are at home or away. Astro also interacts with Amazon's Ring security system, boosting its capabilities to include doorbell monitoring and alarm response.

3. Personal Assistance: Astro serves as a personal assistant, answering commands and requests using Amazon's Alexa voice assistant. Astro can help you with appointment reminders, weather updates, and control over your smart home gadgets. It can follow you throughout the house to ensure that Alexa is always within reach.

4. Entertainment and Communication: Beyond its practical uses, Astro also provides entertainment. The built-in display and speakers allow the robot to play music, stream videos, and even make video conversations. Its portability enables you to watch movies and stay in touch with loved ones from anyplace in your home.

5. Health and Wellness Monitoring: Astro provides health and wellness monitoring capabilities to households with elderly people or those requiring special care. It can monitor the health status of individuals, remind them to take their medications, and maintain their overall wellbeing by tracking daily routines and alerting caretakers if something appears wrong.

6. Customizable Routines and Integrations: Customizable routines and interfaces with other smart home appliances allow Astro's usefulness to be further expanded. You can program Astro to conduct certain actions at specified times, such as shutting off lights, locking doors, or adjusting the thermostat.

7. Privacy-enhancing features: If you're worried about privacy, you can always establish out-of-bounds zones to indicate areas you don't want Astro to go. There is even a "do not disturb" option accessible. Astro also has AI processors, which enable speech and edge computing for a variety of applications. In theory, this means that Astro will analyze a significant percentage of your data locally rather than transferring it to Amazon's cloud, where it can be accessed and securely kept.

8. Compartment Area: Astro features a compartment area that can hold up to 4.4 pounds (2 kilograms) of items. It also has a 15-watt USB-C port that can charge your phone.

9. Package Delivery: Astro revolutionizes package delivery by autonomously transporting small objects to selected spots within your home. Astro's secure compartment allows you to receive packages without worrying about missing deliveries or theft.

10. Play: If you have children, Astro can be a very appealing (although pricey) toy. Although Alexa is on board, Astro has a distinct personality. You can ask Astro to dance, beatbox, rap, burp, imitate various animals, speak, sing, and do a variety of other things, and each request will be met with adorable movement, expression, and sound.

Unboxing Your Amazon Astro

The first step in integrating Astro into your house is unpacking the robot. Carefully open the package to confirm that all required components are there. These might include the robot, charging connections, attachments, and any other equipment you've purchased. To unbox the Amazon Astro, first remove the side clips labeled one and two. These clips seal the wrapping, and once removed, the box becomes ready for opening. The package contains a variety of materials, including stickers, a welcome brochure, and the charging station. The charging station is hefty and has cable management tools to keep things orderly. It also includes a cup holder and a power outlet at the rear.

The Amazon Astro is a sleek, futuristic-looking robot with wheels on the bottom. It has various sensors and a tiltable screen. The robot's top includes three buttons: a minus sign, a plus sign, and a mute button. Overall, the

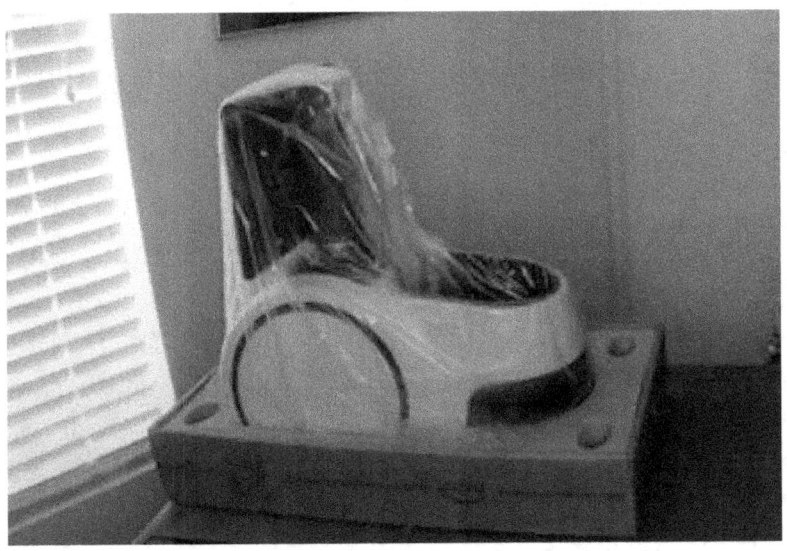

unpacking procedure is simple, with no difficult processes involved.

Setting Up Astro

The unboxing reveals a welcome pack, instructions, and a charging base resembling that of a robot vacuum. Before putting Astro to the test, you would need to set up the accompanying application and synchronize it with the robot. This stage entails prepping the designated regions for Astro's navigation, including closing doors and eliminating small items.

Make sure you have at least 1-2 hours available to finish the basic Astro setup. It will probably take a little more than an hour to train it to recognize your voice/face, investigate your house, and then tour your home and identify every room. Following that, you'll probably want to spend more time exploring with some of its features.

Prerequisites

Here are some planning guidelines to help with the first setup:

- Remove any barriers that may be in Astro's path while he explores your home. For example, any towels or clothing on the floor, loose objects, brooms or mops, and so on.
- Close any doors that you would not wish to include in your home exploration.
- Identify a space in your home that will (semi-)permanently house Astro's charging base. If you relocate it after installation, Astro will have to re-

explore your property. As a result, determine where it will be charged ahead of time.

How do you set up Astro ?

- The directions for unwrapping Astro are included on the white label next to the box's clips.
- The next step in getting the Amazon Astro ready for usage is to set up the charging station. The charging station should be put in an appropriate area, and the supplied menu has useful advice on how to do so. Once you have completely set up the charging station, connect the power cord and utilize the cable management tools to keep things tidy. The Amazon Astro includes two huge rubber feet on the bottom to ensure that it remains in place and does not slide. Astro will remember this place and return for charging. Please ensure that this space remains clear of obstacles. With the charging station in place, the Amazon Astro is now ready for you to charge and use. For Astro's charger, place the charging dock on a flat surface near a wall outlet that is approximately 3 feet wide by 6 feet long. Put your Astro directly in front of the charging port, screen facing away from it. Carefully roll it forward onto the charger.
- Astro turns on.

If Astro does not turn on, push and hold the microphone/camera off button for 2-3 seconds until it does.

- Astro will then search for networks, choose your WIFI SSID (network name), and input your password.

- You will then be invited to check in to your Amazon account, enter your email or phone number and password, and then click the CONTINUE option.
- Once registered, select the CONTINUE option.
- Read the notice and then hit AGREE AND CONTINUE.
- Select your time zone and then press the CONTINUE button. Then, continue with the MEETING ASTRO SETUP and take a few steps back.
- The next step is for your Astro to undock and ask you to perform a few tasks, such as making it dance

or flip a coin.

- Next, Astro will want to construct a voice and picture profile for you. Choose your profile name from the list.

- To teach Astro to recognize your voice, click on AGREE & CONTINUE and follow the on-screen directions.
- Click on AGREE & CONTINUE to train Astro with a visual ID. This will enable it to identify you in your house.
- Follow the vocal cues supplied by moving your head to the appropriate position. Repeat the process with any more family members. If they are not already accessible, you may add them later by swiping down from the top of the screen and selecting Profiles.
- Download and set up the Amazon Astro application. To pair Astro with your account, use the Astro App and scan the QR code on its display.
- Astro will now practice docking; give it a few minutes.

When it's ready, touch the button to let Astro start exploring your house. You could decide to follow it around to ensure it does not become trapped anywhere. Generally, it should do this on its own. It may take more than 20 minutes for completion and will return itself to the docking station once finished.

- Astro is now prepared for a home tour. Tap Continue; during this phase, walk Astro to the center of each room. Then say: "Astro, this is the

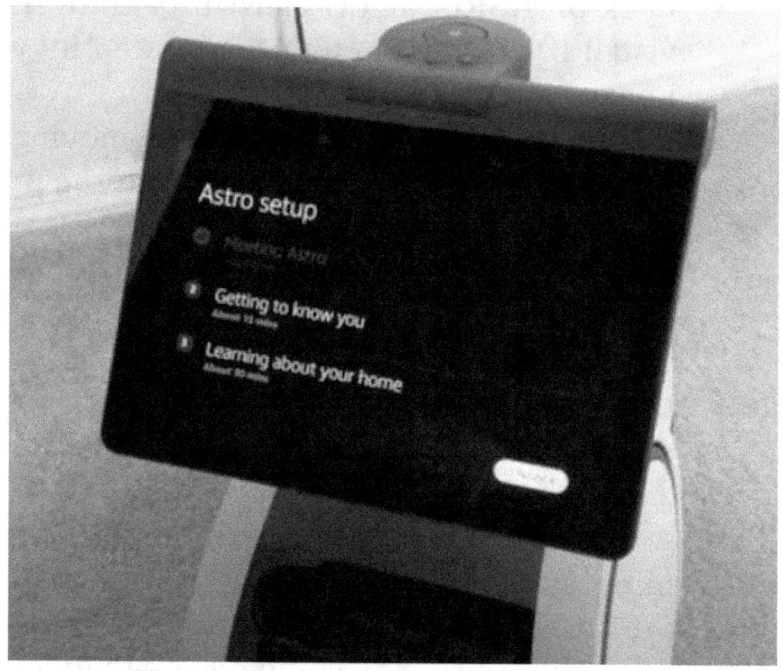

[room name]". Repeat for each room in your home.

- Astro will create a map of your house that can be viewed via the Astro App.
- Once all rooms have been recognized, the setup is finished!

Astro Application

The Amazon Astro app is available on both the Google Play Store and the Apple App Store. This application may be installed at any time, whether before Astro arrives or after Astro has been put on the charging station. However, you will not be able to add your Astro to the app unless it is first charged and linked to your WIFI network.

The Astro application has four (4) tabs down the bottom of the screen. Each of these tabs will be discussed further below.

1. Astro

The Astro view is the first tab located at the bottom of the screen. It gives a summary of Astro's condition, battery capacity (in percentages), history, and events.

Home Monitoring

At the upper part of the Astro tab, lies three (3) Home Monitoring choices. They are:

- Disarmed - When Astro is set to Disarmed, Astro will not conduct any monitoring actions.
- Home: Astro will patrol your house using the frequency you designate in Settings → General→ Home Monitoring → Patrols → Patrol frequency in Home mode (you can modify this frequency by clicking on Edit).
- Away: This indicates that you are Away from Home. When in Away mode, Astro will patrol your house using the frequency you choose in Settings → General → Home Monitoring → Patrols → Patrol

frequency in Away mode (you may modify this frequency by clicking on Edit).

Status Information

The current position of Astro in your house is displayed underneath Home Monitoring. There will also be a battery indicator that will show you the current charge condition of the battery.

Below the battery indication, you'll see a blue sign with a map of your house layout underneath, showing Astro's present position.

2. Live View.

Astro will be able to communicate with you from inside your home through the Live View tab. After selecting the Live View option (at the bottom of the app), anyone near Astro will hear a sound indicating that you are remotely seeing Astro's cameras (tele-presence).

The options available at the top allow you to execute the following actions:

- Stop View - Turns off the Live View; the live video feed from Astro's camera(s) will no longer be sent to your smartphone.

- Go To - This is a tab that you will most likely utilize frequently. It allows you to deploy Astro to a certain room in your home by just pressing on it. There are several extra settings here that you'll certainly find quite useful as you use your Astro more regularly, and these are:

- ROOMS - Select a room from the list; you will then see the live video feed change and Astro go

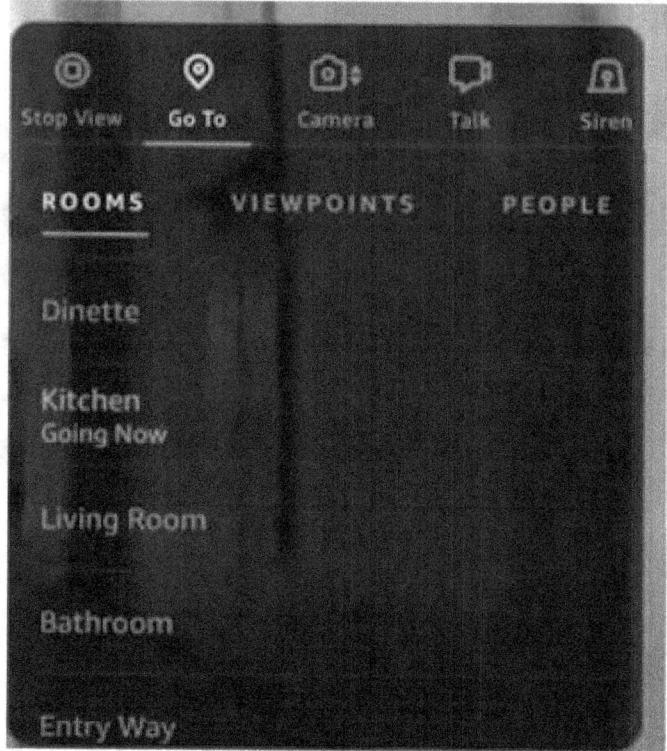

to that room.

- VIEWPOINTS - Viewpoints are very helpful. Viewpoints not only direct Astro to a specific spot in a room, in addition it can also expand the Periscope for greater viewing. Here's how:
  - Navigate Astro to the desired room (while in full-screen mode).
  - Astro can be moved forward or backward by clicking on the up/down symbol located in the lower-left corner of the screen. Swipe left/right to go

right/left. Swiping up/down will extend/reduce the Periscope.
- When you've finished customizing your view, collapse the full-screen view (lower-right icon).
- Tap the CREATE a new viewpoint option and give this viewpoint a memorable name (e.g., Christmas Tree, Livingroom Sofa, Cat House, etc.).
- If you want your Astro to return to this Viewpoint, simply tap the Viewpoint, and it will do so. To delete the Viewpoint, swipe from right to left and hit Edit. At the bottom, tap REMOVE VIEWPOINT.

o PEOPLE - This tab will show you when Astro last saw a member of the family.
o Camera - Three options are available under Camera.
- Video- Any video you've captured will be displayed in the far-left icon. The center red record icon allows you to record with Astro's camera. The flashlight icon will activate a light for a better view in poorly lit locations.
- Photo - The far-left icon displays any images you've already taken. The center white button will take a still image of the current video feed delivered to your smartphone.

> The flashlight icon will activate a light for a better view in dimly lit locations.
> - Adjust - There are two sliders accessible here: periscope and display. Touching the Display slider up or down adjusts the angle of Astro's display. Tapping on one of the five dots under Periscope will raise it to the desired level.
>   - Talk - This option allows you to have a video or an audio chat with anyone nearby Astro. It offers a true telepresence mode of communication. To enable or disable any of the functions listed here, tap the Mic, Video, or Sound icons.
>   - Siren - From the Siren tab, you could click on the Start Siren button to play a siren for 30 seconds. I'd only recommend doing this in an emergency or if you can't reach someone via the Talk option. The siren is rather loud, which may scare anyone around.

3.Settings

Clicking the Gear or Settings icon at the bottom of the Astro app will display a variety of options for customizing its configuration, functioning, and interactions with you and your family. The essential options and their functions will be covered in detail below.

Devices

Astro device(s) that are accessible through your account.

There are the following choices available on those devices:
- Name - The name most likely given to your Astro device; however, it may change once allocated to your account.
    - Pair with Astro - Allows you to link Astro to your account. To pair, both your device (phone or tablet) and Astro must be connected to the same WIFI network. You could click on the the Scan QR Code, Generate PIN Code, and Skip buttons. The simplest technique for me was to select the Scan QR Code option, aim my phone at the QR code on the Astro display, and pair it with Astro.
    - Wake Words – You can choose from any of the many options as your wake word, such as "Astro" and "Alexa", "Astro" and "Amazon", and so on. Choose the one you like.
    - Battery - Simply displays the Astro battery's charge percentage.
    - Behavior – Behavioral selections for Astro.
        - Do Not Disturb - If this is activated, Astro will not disturb you except to give timers, alerts, or reminders that you have previously set.
        - Schedule - You can choose to set a daily schedule for when DnD will be activated to reduce interruptions.
    - Restart Astro - This function allows you to restart Astro in the same way that you would power off and on.

- Factory Reset - This option resets Astro as if it came from the factory. If you execute a factory reset, you must restart the whole setup procedure. This is important if you subsequently decide to sell your Astro or have a severe problem that cannot be fixed otherwise.
- About - This will display the Device Software Version and Serial Number for your Astro device.

Map

Astro generates a map as it explores your house. This map is kept on Amazon cloud servers utilizing AWS (Amazon Web Services), the platform utilized by Astro, Alexa, and the majority of Amazon cloud-based services. The Map option has the following options:
- Edit Map: By selecting Edit Map, you may change the Edit Out of Bounds option and make changes to

individual rooms. Once Astro explores your house, you will most likely want to mark certain locations as off limits. Particularly in toilets and other locations where you don't want Astro to intrude. After selecting the Edit Out of Bounds option, create an orange box around the regions where Astro cannot access.
- Rotate - Use the Rotate option to rotate the map to the orientation that appears to make the most sense to you. You must then click on the Save button in order to save the rotation setting alteration.

General

This option's settings setup is worth taking a good look at. There are several choices available to make operating Astro easier and enhance your overall experience with the robot.

Astro may issue Smart Alerts if it recognizes specific occurrences in your house. We will briefly discuss each of the primary possibilities below.

Smart alerts

Smart Alerts are divided into three (3) major categories: Astro Alerts, Alexa Guard Alerts (Away Only), and Ring Alerts. Let us discuss each:
- Astro Alerts - This feature can identify unrecognized individuals in your house while in the Home or Away mode. Enable or disable these settings based on your preferences.
- Alexa Guard Alerts (Away Only) - You can set up alerts to be alerted and recorded when a smoke or

CO alarm sounds, a glass break is detected, or activity sounds.
- Ring Alerts - Astro can analyze Ring-detected alert occurrences. For example, if you have a Ring doorbell, this option will allow Astro to investigate the scene of the occurrence.

Patrols

Astro, while in Away or Home mode, may be programmed to patrol your house on a regular basis and warn you of any detected activity. These include:

- Patrol frequency in Away mode - When Astro is in away mode, you may configure it to patrol every Never, 1, 2, 4, 6, 12, or 24 hours.
- Patrol frequency in Home mode - When Astro is in away mode, you may configure it to patrol every Never, 1, 2, 4, 6, 12, or 24 hours.

Alexa Guard

This allows you to adjust, edit, or remove Alexa Guard Settings.

Ring

This identifies if you have the associated Ring Protect Pro (or Trial) associated with your Astro account. With the purchase of Astro, you will receive 6 months free. You may activate or deactivate the trial at any moment.

Notifications

You may select whether or not to get notifications to your connected device if Astro identifies anything of interest. You'll need to allow or disable those notifications in your phone's settings. When activated, you will get alerts. A few more alternatives are shown below:

- Device Failure - If Astro fails, you may activate alerts by switching this switch to enabled.
- Home Monitoring - You can get notifications when Astro enters Home or Away mode. See the Home Monitoring section for further information.

Video Settings:
This option presently has two possible options, which are:
- Single touch to live view (Enabled/Disabled) - When enabled, the live view starts anytime you tap the live view tab (rather than being turned off by default).
- Invert swipe to turn - I strongly advise you to enable this setting. The default is deactivated, which makes it difficult to get around Astro. Swiping left directs the camera to the right. However, with this option selected, swiping left also moves the camera (Astro) to the left.

Help and Feedback
Various topics and online help for your Astro gadget. This includes Astro setup, Astro home monitoring, contacting Amazon support, and submitting feedback.

Legal
In this tab, you can View and read Amazon's privacy policies. It's a great idea to look over them to have a better grasp of when Astro is recording, how it manages those recordings, and how you may manually remove them.

About the app
Displays the application's version.

Sign Out
This option allows you to sign out fully from the Astro application. After selecting this option, use the OK button to sign out.

4. Voice
Think of the Voice tab as a walkie-talkie for Astro. This option allows you to accomplish a certain operation regardless of his location, exactly as if he were nearby. This is useful if it is in another room in your house; simply say "Astro, go to the [room name]" to call it to the room where you are now located. Astro will then navigate to the room you have specified on its own. The Voice tab allows you to communicate directly with Astro.

On-Device settings
In addition to the settings stated above in the Astro app, there are several settings available on the device itself. To access these options, say "Astro, show settings" or swipe down straight on Astro's touch screen. The following are examples of the main alternatives available:

Bluetooth
Enables Astro to be paired with any Bluetooth-capable device. Tap your smartphone to pair.

Brightness
This adjusts the brightness of Astro's display. When the Adaptive Brightness option is activated, it will

automatically adapt based on the existing light.

Sounds
This helps you adjust sound levels for any media playing, equalizers, timers, alarms, and notifications. You may also set or disable Ascending Alarms, it will progressively raise to your desired level. Under the Driving selection, you may modify the driving sound to Tones (mild humming), None (off), or Clicks You can also activate or disable noises at the beginning and end of a vocal request.

Amazon Kids
This feature is designed specifically for families with children aged 13 and younger, and it may be enabled or disabled. It will block the playing of adult-oriented content.

Do Not Disturb
This disables alerts, there's an exception for alarms and timers, incoming live view requests, and autonomous happenings (patrols and hanging around). You may also arrange when Do Not Disturb is activated on a daily basis by specifying a Start and End time.

Cameras & Live View
- Streaming Cameras: If this option is selected, video streaming functions and photo functions (including video calls, photography, and recording) will not function while Astro's periscope cameras and display are off.

- Alexa and all Other Amazon Devices - Allows or prevents remote access to Astro's camera stream via compatible Alexa devices associated to this account.
- Astro App - Simply indicates that there are additional possibilities with the Astro application for your phone.
- Paired Mobile Devices - Displays all the mobile devices that have been paired with Astro.
- Streaming Delay - Sets a delay before the live view begins, except while in away mode.

Communications
- Incoming Call Ringer – Disable or Enable this option.
- Drop In: Disable or enable this feature. The Drop In function enables household members as well as contacts to easily connect directly to the Astro device.

Device Options

- Device Name - The name of the Astro device.
- Device Location - shows the address and exactly where Astro is located.
- Wake Word - Change the wake word being used by Astro.
- Date and Time - Adjust the date and time.
- Temperature - Change the temperature reporting mechanism.

- Distance - Change the distance reporting technique.
- Web Options - Settings for the web browser (Silk) and online videos.
- Night Mode Options - Night mode settings.
- Hangout- Astro will occasionally put down the charger to spend time with family members. If you don't want Astro to move on its own, you may turn

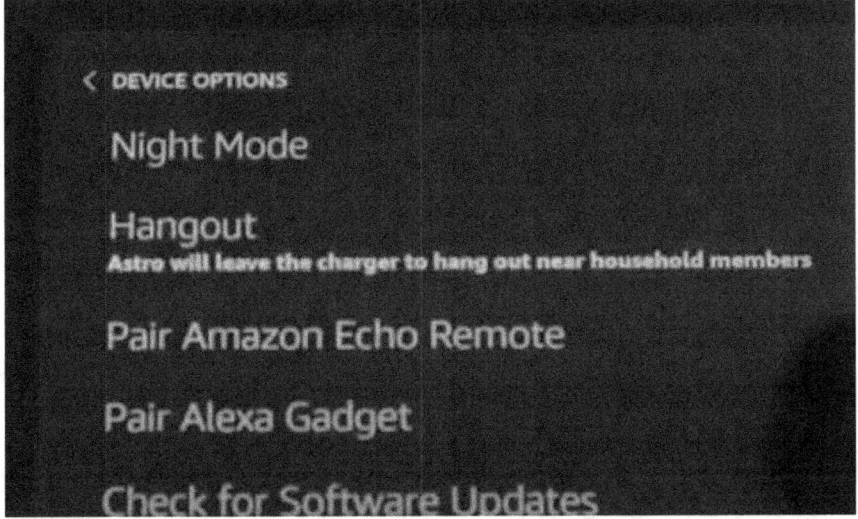

this function off.

- Extra Data Collection for Drop In - Enable or disable for more information to be recorded.
- Pair an Amazon Echo Remote - Connect an Echo remote to your Astro.
- Pair Alexa Gadget - Connects an Alexa-enabled device to Astro.
- Check for any Software Updates - This feature allows you to check for new software updates and see the current operating system version.
- Serial Number - Astro's serial number.

Map Setup
After authenticating your account, you may change the map settings.

Profiles
By choosing one of the above profiles, you may control the voice and visual identification for the specified family members.

Restrict Access
- Amazon photos - Disables search and no more uses photographs as the home screen backgrounds or displays Hoto Highlights on your home screen.
- Movie Trailers - Use blocks to search for things like movie trailers.
- Web Browser – This will disable web searching and surfing.
- Web Video Search - Prevents voice searches or restricts the results.
- Video Providers - Block or restrict the video sources that are enabled on this device.

The Design and Hardware Breakdown

Amazon Astro, a house robot that will be released in September 2021, is the company's bold entry into the consumer robotics market. Astro is designed to integrate into smart homes and perform a range of duties, including home monitoring, task assistance, and functioning as a smart assistant, all while communicating with users in an engaging and friendly manner. Its design and hardware reflect years of advancements in robotics, human-computer interaction, and artificial intelligence. This breakdown will go over Astro's physical features, sensors, and capabilities, demonstrating how its design contributes to it being an intuitive, user-friendly, and valuable companion in the house.

Physical design.
The Amazon Astro's design seeks to find a balance between practicality and friendliness. With a small, rolling body and a screen for a face that combines functionality and approachability, the robot looks like an upgraded version of a vacuum robot.

Size & Dimensions:
Astro is compact and mobile, and its size make it easy to explore common residential surroundings. The robot is roughly 17 inches tall, 9.8 inches wide, and 16.7 inches deep. It weighs around 20 pounds, which is substantial enough to give stability while remaining light enough to allow for safe mobility over diverse types of floors.
Astro's size and weight are carefully considered to ensure that it can pass through regular doorways, move around furniture, and not seem overbearing in smaller rooms. This tiny shape is especially useful for traveling through confined locations and connecting with consumers in a functional and welcoming way.

Wheels & Mobility

Astro's mobility is one of its most distinguishing traits. It has two huge driving wheels and a smaller, omnidirectional caster wheel for stability and ease of use. Together with the caster wheel providing balance and accuracy while turning and maneuvering around tight corners, the huge wheels let it to travel effortlessly across a variety of surfaces, including tiles, hardwood floors, and carpets. The robot's capacity to navigate complicated household surroundings is made possible by powerful algorithms that integrate mapping, obstacle avoidance, and real-time decision making. It moves smoothly and quietly, making it ideal for use in the house. Astro may also change its speed according on its environment, going slower in busy areas and quicker in open regions.

Adjustable Periscope Camera
Astro's design distinguishes itself with its periscope camera. This camera may stretch upwards from the body to get a better view of the surroundings. The periscope can elevate approximately 42 inches measuring from the floor, making it ideal for inspecting items that are above Astro's regular field of sight, such as worktops or stove tops. The camera may spin and tilt, providing full visual coverage of a room or region.

This capability enhances Astro's home monitoring capabilities by allowing it to visually investigate regions that a stationary camera or low-to-the-ground device may overlook. The periscope camera boasts an excellent 12 MP quality, giving customers with crisp and vivid pictures while watching their house remotely.

Screen and Face

Astro's "face" is a touchscreen display situated in front of the robot's torso. This screen serves both a practical and aesthetic purpose. Functionally, it shows critical information, and notifications, and lets users interact with it through touch commands.

Aesthetically, the screen is intended to replicate the emotive quality of a face, therefore humanizing the robot and making interactions much more engaging and less

mechanical.

The face has basic, cartoonish eyes that can blink, glance around, and exhibit different emotions depending on how it will interact with people.

This degree of expressiveness is intended to make Astro more approachable and less scary. Astro's eyes move in reaction to orders or requests, and they may even represent his attitude or state, such as displaying a neutral expression when idle or a cheerful smile when conversing with another user. This contributes to its affable and welcoming personality.

Compact form factor

Astro's small form allows for easy movement throughout the home. Its rounded corners and smooth, white plastic body are purposely minimalistic, giving it a clean and modern appearance that blends in with any decor without drawing attention to itself. The robot lacks arms and sophisticated appendages, which simplifies its construction while also making it more robust and less susceptible to mechanical problems.

The top of the robot has a small storage box that may be used to transport lightweight objects about the house, such as a can of soda or small gifts. This innovation, while not novel, adds a level of ease and utility to the Astro design.

Sensors & Navigation

Numerous sensors included in Astro's hardware design enable safe and efficient movement and interaction with its environment. Astro's ability to map its surroundings, identify obstacles, and communicate with people depends on these sensors.

Navigation Sensors

Astro uses a range of powerful sensors and mapping tools to move around dwellings on its own. The primary sensors are:

• Light Detection and Ranging, or LIDAR: Astro creates a detailed three-dimensional map of its environment using LIDAR technology. The way this gadget operates is by

pulsating laser light and timing how long it takes for objects to bounce back after impact. Because of this, Astro can accurately map a room's layout, including the locations of furniture, walls, and other obstacles. Astro's autonomous navigation technology is based on LIDAR, which enables it to detect and avoid obstacles and also detect the best routes across a house.
- Ultrasonic Sensors: In addition to LIDAR, Astro also possesses ultrasonic sensors, which help with close-up item and person detection. In order for these sensors to function, high-frequency sound waves are produced, and their return time after bouncing off objects is tracked. This allows Astro to recognize smaller or softer items that LIDAR may miss, such as drapes or little dogs.
- Depth Sensors: Astro's distance from nearby objects is measured using these sensors. These sensors are essential to preventing Astro from running into walls or furniture. Additionally, Astro may navigate around dynamic obstacles like people or pets using its depth sensors. These objects may move suddenly.
- Bump Sensors: Astro's front body is equipped with bump sensors. These sensors are designed to detect inadvertent collisions and notify Astro to reroute and stop. The bump sensors act as a safety measure even though Astro's better sensors often prevent collisions.

Cliff Detection Sensors: Astro can distinguish between a ledge and a stairway thanks to cliff detection sensors on its underside. These sensors are essential to ensuring Astro's safety in multi-story homes by preventing it from inadvertently sliding down stairs or other drops.

Cameras for the purpose of visual recognition.
Numerous high-caliber cameras fuel Astro's sight recognition system. These cameras may be used for many different things, such as home security and family member identification and communication.
- Front-facing Camera: Astro's main camera is placed above the screen and is in charge of facial recognition and video calls. This camera has a 1080p resolution and is utilized for video conferencing, recognizing household members, and taking photos while monitoring the home.
- Periscope Camera: The aforementioned periscope camera provides a distinct benefit for monitoring places beyond Astro's typical range of sight. Its capacity to raise and rotate allows you to remotely evaluate different portions of your home, such as if

the stove is on or a window is open.

Astro's cameras, paired with powerful machine learning algorithms, enable it to do face recognition, discriminate between various users, and respond appropriately. This allows Astro to be more personalized by greeting consumers by name and remembering their preferences.

Microphones & Voice Recognition

Astro has a set of far-field microphones that allow it to receive voice instructions from across the room. The microphones are positioned in a circular array, similar to those used in Amazon Echo devices, and enable full 360-degree voice recognition coverage. These microphones enable Astro to reply to Alexa voice commands, answer questions, operate smart home devices, and engage with people in a natural, conversational manner. The speech recognition technology is intended to eliminate background noise and focus on the individual issuing orders. This enables Astro to interpret commands even in loud surroundings, such as a kitchen with running appliances or a living room with the television on.

Functionalities and Use Cases:
Astro's hardware supports a wide range of functions aimed at improving the smart home experience. Its capacity to walk independently, identify people, listen to spoken instructions, and interact with its surroundings makes it an adaptable tool for home surveillance, communication, and entertainment.

Home Monitoring

One of Astro's key responsibilities is house surveillance. Astro, with its cameras and sensors, can patrol the home while the users are away, delivering real-time video feeds and alarms. If the robot senses odd behavior, such as an unexpected door opening, it may notify the owner by smartphone.

Astro may be set to follow certain patrol routes or inspect specific locations at predetermined intervals. This feature is especially beneficial for individuals who wish to monitor pets, children, or valuables when they are away from home.

Alexa Integration

Astro, like the rest of the Amazon Echo line, has Alexa integrated in. This enables users to ask Astro questions, set reminders, operate smart home gadgets, and even conduct video chats using voice commands. Astro's mobility enhances the Alexa experience by allowing the robot to accompany users from one room to room, delivering hands-free help as needed.

Communication & Entertainment

Astro's video calling capabilities are another important feature. The robot's screen can be utilized to carry out video chats, and its ability to move allows it to follow people during calls, keeping them in frame. This makes Astro an excellent tool for staying in touch with family members or acquaintances, particularly for those who live alone or have mobility issues. Astro can not only communicate but also entertain. It can play music, show videos, and display information on its screen. Astro's interactive personality brings a fun, engaging aspect to these activities, making it feel more like a friend than a standard technology.

Battery Life and Charging

Astro is equipped with a rechargeable lithium-ion battery. It can run for around two hours on just one charge, depending on how it is used. When the battery runs low, Astro will automatically return to its charging dock to refuel, ensuring that it is always ready to go when required.

The charging port is small and meant to be put in a low-traffic area of the house so Astro can readily use it without getting in the way. The robot docks by utilizing sensors that align with the charging connections, which takes a few seconds.

Privacy Features

Given the significant usage of cameras and microphones, privacy is a big problem for any house robot. Amazon has addressed this by including many privacy safeguards. Astro has a "Do Not Disturb" option, which turns off the microphones and cameras when the user wants privacy. Furthermore, a physical button on the robot's torso allows users to disable these sensors at any moment. Astro also has visual and audible indications to help users know whether it is recording or streaming video. These capabilities provide customers complete choice over when and how the robot collects data, solving privacy concerns while keeping fundamental functionality.

Amazon Astro's design and hardware strike a delicate blend of usefulness, usability, and technical complexity. Its tiny, accessible form, along with powerful sensors and navigation algorithms, make it a useful and fascinating addition to any modern smart home. Astro, with features ranging from home surveillance to interactive entertainment, provides a look into the future of robots in daily life. While the technology is still maturing, Astro exemplifies Amazon's dedication to combining robots and AI to build useful, intuitive, and responsive home companions.

How Astro Connects with Current Smart Home Devices and Alexa

Imagine a time when the smart assistant in your house follows you about, anticipates your needs, and engages with your environment in real time—rather than merely responding from a fixed gadget on a shelf. Presenting Amazon Astro, a mobile version of Alexa that elevates home automation. This study is for you if you've been wondering how Astro works with Alexa and the smart home appliances you already have.

Astro provides a novel perspective on what home assistants can accomplish by fusing cutting-edge technology, sensors, and artificial intelligence with Alexa's vocal skills. Essentially, Astro makes Alexa more than just a static interface; rather, it makes Alexa a dynamic, portable, and incredibly responsive element of your everyday life.

Astro: A Smart Hub on the Go

The fundamental feature of Astro's Alexa integration is its capacity to serve as a mobile center for your smart home. Due to its mobility, Astro can follow you around and respond to voice instructions without any need for fixed smart speakers or screens. Astro can track and react in real-time whether you're traveling from room to room or cooking in the kitchen or lounging in the living room. Because Astro has all of Alexa's capabilities, you can ask it to do everything from lock your doors to control your lights to change the temperature. Astro seamlessly integrates into homes with pre-existing smart ecosystems since Alexa is already compatible with a large number of smart home products. By adding a layer of mobility and autonomy, Astro enhances the experience.

How Alexa and Astro Work Together: Smooth Voice Control

Astro received all of Alexa's voice control features due to its connection with the artificial intelligence assistant. Whether you're new to smart home technology or a seasoned Alexa user, you'll discover that Astro reacts to voice commands in the same manner as any other Alexa device.

Voice Activation Throughout Your House

Being able to call upon Astro from anywhere is among its most appealing characteristics. When using a standard Alexa device, you have to be able to see the speaker or display in question. However, Astro's mobility modifies that. Say "Astro, come here," and the robot will find your position on its own and bring Alexa's features with it. Larger homes with numerous stories and rooms can benefit greatly from this feature.

Envision yourself in your bedroom, seeing that the lights in the living room are still on. Call Astro, tell it to turn off the lights, and you can accomplish the task without getting out of bed. This is a more convenient option than going to the closest Echo speaker.

Dynamic Interactions

The capability of actively interacting with your surroundings is another benefit of Astro's connection with Alexa. Conventional Alexa devices provide a fixed point of engagement since they are stationary. Conversely, Astro has the ability to walk around your house, facilitating more organic connections. To actually ensure that you maintain control regardless of where you are located in your house, Astro can move around with you as it accomplishes tasks like locking the front door when you go upstairs.

Because of its mobility, Astro can also provide assistance proactively. For example, Astro can come over to you to validate the operation and provide feedback if you say, "Alexa, turn off the oven," while laying the table for dinner. Because Astro can respond to queries and offer a more engaging experience than a fixed gadget could, conversations become more personal as a result.

Smart Home Appliances with Astro: Complete Integration with Your Current Setup

One of Astro's greatest qualities is how well it integrates with current smart home appliances. Astro can communicate with and manage your whole collection of smart lights, thermostats, security cameras, and other Internet of Things (IoT) devices with the same ease as any Echo device.

Compatibility Among Devices

Being compatible with the same spectrum of smart home devices that Alexa can manage, Astro is very compatible with the whole Alexa ecosystem. Astro can instantaneously connect to any Alexa enabled device you currently own, including smart locks, plugs, and other accessories. For instance, you may ask Astro to alter the lighting as it follows you from room to room if your smart lights are linked to Alexa.

You may keep using your current smart devices without having to change them because Astro utilizes the same

Alexa platform. Astro immediately acquires access to all of the devices that are already connected to Alexa once it is paired with your Alexa account and connects to your home Wi-Fi. Because of this, integration is simple and painless, needing little more preparation beyond the first setting.

Monitoring and Security for Your Home
Astro's capacity to serve as a mobile security monitor is among its most intriguing aspects. Astro can automatically patrol your house, identify strange behavior, and instantly warn you thanks to its powerful AI, cameras, and sensors. Astro easily pairs with Alexa Guard, a security feature that watches for warning indications of danger, such as the noise of a smoke alarm or shattered glass, and is included into Alexa devices. Combining this with Astro's mobility and AI-powered vision results in an even more potent home security system. Through the Alexa app, Astro can physically investigate different regions of your house and provide live video feeds to your phone. You may instruct Astro to check on particular rooms or make sure doors and windows are locked whether you're at work or on vacation.

Unlike other fixed cameras, Astro's periscope camera offers a unique viewpoint that enables it to see objects from a higher perspective. This is especially helpful for keeping an eye on places that are difficult to get to, such upper windows or high shelves.

Astro may also be configured to communicate with other smart security systems in your house. You could program Astro to arm the security system, lock all the doors at night, and then patrol the house. Astro can look into the matter right away and deliver a live video feed to your phone if a smart motion detection goes off.

While you're away, Astro watches over your area. Astro

can also be used for:

- Use the Astro app to enable you see your space in real time.
- Get notifications when someone unfamiliar enters your area that Astro has detected.

Astro could do even more if you have linked your Ring and Amazon accounts and are a Ring Protect member. Astro can be set up to:

- Patrol your area and, if you'd want, use Ring's cloud storage to stream and also save a video from every scan point Astro visits.
- Examine any events that have been noticed. Accompany an unknown individual. Ring's cloud storage allows you to stream and save video.

How to Set up Amazon Astro Monitoring.

To configure monitoring on Astro, use the Astro app.
For setting up Astro for monitoring:
1. Launch the app for Amazon Astro.
2. Click on Settings.
3. Select the option Home Monitoring.
Here, you may set up Astro for monitoring in several ways, including:
- Patrol (each having a paired Ring Protect account)
- Voice Guard
- Ring (associated with a paired account for Ring Protect)

You may instruct Astro to monitor your area by using the following commands once your monitoring features have been configured:
- "Astro, I'm leaving." When you're about to go out.
- "Astro, set up monitoring to Home." - While you're at home.

Turn off the Amazon Astro Monitoring system.

Follow these instructions to turn off monitoring using the Astro app or your voice.
Tips: Using your voice to disarm Amazon Astro requires a

Ring Protect Pro membership. Choose Settings > Home Monitoring > Disarm Monitoring by Voice in the Astro application to "on" the Disarm by Voice option. Say, "Astro, turn off home monitoring," to disarm. Astro will ask you for your code to disarm if you have one.
Using the Astro app, disengage Amazon Astro monitoring:
1. Launch the Amazon Astro App.
2. Select Astro.
3. Click on Disarm.
Note: You must "on" the Ring skill in the Alexa app in order for Disarm by Voice to function with Ring alerts. Go to Turn Alexa Skills On or Off for additional details.

How to use Multiple Amazon Astros to Monitor Your Space.

Every Amazon Astro gadget has to have its own charging port and be able to monitor a certain region. Multiple Astro devices cannot be used in the same space. Note: Astro's Home and Away security modes are only supported by the Alexa app for one Ring location. Add your Amazon Astro device to each location in the Ring app so you can monitor multiple Ring locations. Next, use the Alexa application to turn "off" the Astro device modes.

Personal Assistance and the customization option on Astro

With Astro's Alexa integration, you can do more with your smart home than just use it as a personal assistant because of its AI capabilities. Because of this feature, Astro becomes more than simply a tool you control—rather, it becomes a lifelong friend that gets to know you and your routines.

Enable and Disable home monitoring using modes
What are modes?
Modes are Astro app options that regulate Astro's Home Monitoring functions while you are at home or away. There are three modes for controlling the condition of your house: home, disarmed, and away. For example, set Home Monitoring to Away mode to receive an app notice while Astro is on patrol and identifies an unknown individual. If you're staying in for the night, you may set Astro to 'Disarmed'.
Astro Modes may also be synced with compatible Ring devices (doorbells, cameras, and Ring Alarm, each sold separately) if you have a Ring Protect subscription.
As of March 29, 2023, a Ring Protect membership is needed in order to sync Modes between Astro and your Ring devices, if you have any. You can utilize Modes with Astro even without a Ring Protect membership, but a Ring Protect subscription is needed for Astro and Ring devices to sync automatically. When you switch Astro's mode (to Disarmed, Home, or Away) without having a Ring Protect membership, the other Ring devices' modes stay the same and don't sync with the mode you've chosen on Astro.

Setting Up a Smart Alert on Amazon Astro
Configure recordings or smart alert alerts for when Astro notices something like an unfamiliar individual.
Use Smart Alerts to get alerts when Astro notices specific occurrences, like:
- An unknown individual
- The alarm sounding for CO or smoke
- The sound of glass shattering

- Anything that sets off the sensors on your Ring Alarm.

To configure or update Smart Alerts:
- Launch the app for Amazon Astro.
- Click on Settings.
- Click on Home Monitoring and adhere to the prompts displayed on the screen.

Tip: After turning on Home or Away mode, cover huge images or posters of humans, mannequins, or sculptures. This prevents Astro from identifying items as unidentified humans.

How to Setup the Ring Protect Pro for Amazon Astro

Begin pairing your Ring Protect Pro account with Astro via the Astro app. If you don't already have it, download the Ring app from your mobile device's app store before you begin. Visit support.ring.com if you need help installing the Ring application.

To connect your Ring Protect Pro account with Astro:
1. Start the Amazon Astro app.
2. Select Settings.
3. Select Home Monitoring.
4. Select Ring.
5. Select Add to Ring App.

The Astro app takes you to the Ring application, where you can sign up for Ring Protect Pro and complete the setup.

Routine Automation

Routines are one of the best ways Astro and Alexa can interact. You may set up routines with Alexa that, when given a single command, cause a chain of events. As an illustration, you may have a "Good Morning" ritual that includes setting the lights to turn on, brewing coffee, and receiving the weather report. Astro goes above and above with these performances, literally performing alongside you.

For example, when you say, "Alexa, start my day," Astro may follow you throughout the house, providing reminders or modifying the setting according to your preferences, in addition to carrying out the preprogrammed requests. To create a completely customized morning experience, Astro can automatically lift your smart blinds as it enters the room if you have them.

Over time, Astro might also pick up on your preferences. If it observes that you always ask for particular chores at specific times of the day—like turning out your lights at bedtime or shutting the doors after dinner—it might recommend automating these actions for even greater ease.

Management of Families and Visitors

Astro provides an additional level of individualized support by being able to distinguish between various family members and visitors. Astro can respond to the particular tastes of each member of your home and greet them by name thanks to its facial recognition skills. For instance, it may provide one person a separate reminder while reminding you of your impending appointments. Additionally, Astro has the capability to control household profiles, which enables it to respond and offer services that are specific to the individual using it. This is especially helpful in houses with many people, since each member may have distinct habits, tastes, and frequently used smart home appliances. Astro can host guests when they are here. It may lead visitors through the house, guiding them to certain areas and assisting with little things like locating the bathroom or turning on the lights. This gives the experience of a smart home a fresh and captivating touch.

A Human Touch: Astro's Characteristics and Social Features

Astro is more than simply a task-oriented robot; it has a personality that enhances the natural and enjoyable experience of engaging with it. To make sure that Astro doesn't come across as a frigid machine, Amazon has taken considerable care. Rather, it is intended to be expressive and participatory, communicating through a mix of noises, motions, and visual displays.

Pleasant and Interesting Interactions

The animated face of Astro is displayed on its display screen and varies according to interaction and context. Astro's expressive responses, such as its excited greetings or bewildered expressions when it doesn't grasp a command, help to humanize it. Astro's motions also add to its personality; for example, it may tilt its "head" to indicate that it is paying close attention or nod in response to instructions.

Astro is made to interact with its surroundings in ways that are similar to how people do.

For instance, Astro may roll over to you with excitement if you call it from another room, suggesting that it is completely ready to assist. You feel more like you're communicating with a helpful friend rather than simply a computer when you engage in this kind of connection.

Integration of Smart Homes in the Future
The way that Astro has integrated with Alexa and other smart home devices is a major advancement in home automation technology. Through the integration of Alexa's voice-activated capabilities with a robot's intelligence and mobility, Amazon has developed a device that provides tailored and dynamic support unlike anything we've seen before. Astro adds a new dimension of ease and interaction to daily life by assisting you with security, managing your smart home, and serving as a mobile personal assistant.
Astro is much more than simply Alexa on wheels as we look to the future; it offers a window into a world where intelligent technology becomes even more individualized, attentive to our needs, and integrated. Astro represents a significant advancement in home automation technology with its amiable demeanor, sophisticated AI powers, and smooth integration into the current smart home ecosystem.

Astro in Action

One of the most important things for humans has always been staying in contact with relatives, no matter how far away they live. In the contemporary world, technology is a major factor in making that simpler. This concept is taken to a whole new level by Amazon Astro, a smart home robot that runs on Alexa and enables customers to stay connected in a more seamless and engaging way. Astro is definitely more than just a smart gadget; it's made to follow you about, attend to your requirements, and even assist you in taking care of your loved ones while you're not there. Suppose you value staying in contact with your family, in that case, Amazon Astro provides a number of options for you to do so, including video chats, message sending, and secure family member monitoring.

Video Calls: Communicate Face-to-Face from Anywhere

Face-to-face contact is one of the most vital ways we stay connected, and video chats have become a standard part of our modern lives. Video calling is made easier and much more entertaining with Amazon Astro, whether you're checking in with loved ones across the house or catching up with relatives across the nation.

 With the rotating, extended periscope camera that comes with Astro, you can make and receive video calls from any place in the home. With its adjustable height, the camera is convenient to use whether you're sitting, standing, or even moving about. Astro can follow you about the kitchen while you cut veggies or strike up a discussion with a loved one over a video conference, providing you more flexibility and making it way easier for you to stay involved throughout the day.

With Astro's integration with Alexa's video calling capabilities, you can easily make a video call to someone who has the Alexa app or an Echo device simply saying, "Astro, call Mom," or "Astro, follow me," to have a conversation with your loved ones while you go about

your house. The gadget will take care of the rest. You are not need to remain still after the call begins. No matter what room you're in, Astro will follow you around just to make sure you're always in the frame of the camera. This is especially useful when you need to keep your hands free, such as when cleaning, folding laundry, or working on an assignment.
You could change the angle of the camera to get a more comfortable view because it is positioned on a periscope. The camera may be adjusted to the ideal height or angle when video calling someone who is taller or shorter than you. Astro's mobility also enables it to follow you around as you walk, keeping you in the screen and improving the naturalness and flow of the video chat by simulating being in the same room as the other person. Astro also allows group video chats, so you can celebrate important occasions like birthdays, holidays, or family announcements in a more personal way by having several family members on the same call. No matter where you are in the home, Astro's mobility guarantees that everyone, especially those with large families, can see you well.

Sending Messages: Keep in Touch Even When You're Busy

Sometimes all it takes to check in with someone or let them know that you have been thinking of them is a brief note. Sending voice or text messages to your family members without having to pick up your phone is made easy with Astro. When you need to stay connected yet are in the middle of something, this feature comes in handy.
Through Alexa's messaging service, Astro can communicate with other devices that have Alexa enabled.

stating, "Astro, send a message to the kids saying, 'Don't forget to finish your homework before dinner,'" is one way to remind your children to finish their schoolwork. Astro will translate the message. It's a simple, hands-free method of speaking with family members who may be in another room or even elsewhere.

The fact that Astro can receive messages adds to how easy this is. Astro can therefore tell you when someone sends you a message while you're working or preparing, and you may listen to the message straight from the robot. Astro serves as your direct line of contact with your family, so you won't need to search for your phone or consult another gadget.

In bigger homes with many floors or rooms where family members may be dispersed, this capability becomes more useful. Astro allows you to communicate more effectively and efficiently by relaying messages directly to certain individuals rather than you having to shout across the house or send easily forgotten text messages. Astro may be a very useful communication tool in homes with older or less tech-savvy family members. With its voice-activated system, you may communicate with Astro simply by speaking, eliminating the need to fiddle with small screens or complex technologies. It serves as a means of keeping everyone informed, regardless of how engaged or busy they may be.

Keeping an Eye on Your Loved Ones: Instant Peace of Mind

One of the greatest things about Astro is that it can assist you in monitoring loved ones to make sure they are safe and doing well. Astro may be your eyes and ears while you're not at home or are merely in another area of the house. By keeping an eye on your loved ones and notifying you of any strange activity, Astro can give you peace of mind. Astro can navigate the house on its own because to its mobility, and its cameras and sensors give it excellent situational awareness. This makes it the ideal tool for keeping an eye on family members, particularly those who might require more attention. Astro can assist you in keeping an eye on your elderly parent or relative without being too invasive, for instance, if they live with you. When they need help, they may phone Astro to come to their location, or you can schedule regular check-ins with them.

Astro's cameras and its sensors can be accessed from your phone via the Alexa app, providing you with real-time information on events at home. This implies that you may still check in to make sure everything is alright even if you're out at work or running errands. Astro may send an alarm to your phone so that you can act right away if it notices strange behavior, such as someone falling or moving in an unexpected way.

Astro may also be used to watch children, giving parents who wish to continue keeping an eye on their children while working or taking care of other responsibilities an additional degree of security. Simply ask Astro to check up on your child while you're in the kitchen, and the robot will show you a live stream of the area where they're playing securely. It's a simple method to make sure your children are secure without needing to be in the exact same room as them all the time.

Astro not only lets you watch in real time, but it also lets you schedule and establish habits. You may set Astro to check up on your elderly parent, for example, at specific times of the day. Astro is a great tool for keeping up daily habits related to health and wellbeing because it can even remind users to take their medicine, drink water, or complete other activities.

Astro's mobility sets it apart from standard security cameras in situations when you need to monitor different regions of the house while you're gone. Unlike other home security systems, which are set in one place, Astro can move about your house to give you a complete picture of everything going on in various areas and even notify you if there's any strange behavior.

Astro may also be used by pet owners to keep an eye on and communicate with their animals while they are gone. When you're not physically there, you can use Astro's camera to check in on your pets, make sure they're acting appropriately, and even have a conversation with them using the device's built-in speakers. This can provide them a sense of security and comfort.

Astro is designed to watch over your house, so you can chat to anyone around by turning on the two-way communication feature. If you're outdoors gardening, you may use this to console your lonely pet or ask your lover to get you a glass of water.

Select Live View after opening the app.
Click on "Start Live View,"
Then click on "Talk."
In addition to these features, you can make announcements, send messages, and connect to another compatible device.
Say something like, "Astro, make an announcement".

Astro's capacity to keep track of loved ones is not restricted to people who live in the same household. You may utilize Astro's capabilities to remain in touch and make sure your family members who live far away are also taking care of themselves. Those who look after older parents who may live independently but yet need occasional check-ins may find this very helpful. Astro's messaging, video calling, and monitoring capabilities make it simple to stay in touch with distant family members. You may give Astro a call to see how they're doing, leave a note to remind them of something essential to do, or keep an eye on their surroundings to make sure they're safe. Even in situations when physical presence is not possible, this degree of engagement fosters a sense of intimacy. Astro provides a secure and independent alternative for senior family members who might want more regular check-ins but don't want a live-in caregiver to disturb them. You can stay in regular contact and keep an eye on their wellbeing using Astro without being intrusive. With the help of its autonomous functions, the robot can quietly check in on them, and you can view live video feeds to ensure everything is running well.

The capacity of Astro to give care at a distance is very helpful for people who lead busy lifestyles or who reside in different states or cities than their loved ones. With Astro, you may maintain the distance that may be required for your own responsibilities while still staying in touch, watching out for their safety, and showing them care.

Some Commands:
- "Astro, call my family."
  Keep up with Astro's group phone calls.

- "Astro, drop in on the kitchen."
  Connect to Astro-capable gadgets in your house instantly.
- "Astro, let everyone know that dinner is ready."
  Send out announcements to all of your devices that support Astro.
- "Astro, call Dad."
  Hands-free audio or video calling is possible.
- "Astro, find my phone."
  When your phone is lost, ask Astro to call it by calling it.
- "Astro, call for help."
- "Astro, send a message to [name]."
- "Astro, video call [name]."
- "Astro, answer the call."
- "Astro, play messages."
- "Astro, are you hungry?" (This command will enable Astro to communicate whether it requires a recharge, maintenance, or anything else to function properly.)

Amazon Astro is a tool that strengthens family bonds in novel and significant ways; it's more than simply a fancy device. Astro's ability to freely move around your house and communicate with family members offers a unique, hands-free solution that extends beyond the capabilities of standard smart gadgets, whether you're making video calls, sending messages, or making sure your loved ones are safe. Astro gives a degree of personal connection that seems more human and dynamic than inanimate objects because of its capacity to go with you you around, give you real-time updates, and check in on loved ones. It transforms technology into a helpful friend that can ease everyday responsibilities, keep you connected to the people who matter most, and offer you the comfort that comes from knowing your loved ones are secure and well taken care of.

Astro Fun & Games

Astro, the intelligent robot from Amazon, is all about enjoyment as much as functionality, efficiency, and monitoring. Although Astro is intended to be a useful tool for around the house, its entertainment possibilities and interactive capabilities demonstrate that it can also add a fun and lively element to your living space. Astro is an excellent companion as she combines cutting-edge technology with a lighthearted disposition, especially when it has to do with keeping people and kids entertained. Astro's entertainment features are dynamic and adaptable, which gradually gives them a more intimate feel. Astro has Personality traits, it's not Just a Machine.

Prior to discussing certain entertainment aspects, it's important to understand that Astro was created with personality in mind. This voice-activated assistant is unlike others since it can convey a variety of feelings and reactions, giving the impression that it is a kind friend rather than a cold, emotionless device. The sounds Astro makes, the way it tilts its head, and its moving eyes are all intended to resemble a live companion. Through its motions and noises, it might elicit curiosity, enthusiasm, or even a playful attitude. Astro and the residents of the home develop a relationship as a good result of this degree of engagement, which sets the foundation for enjoyable and interesting encounters. When Astro is ordered to follow someone, for instance, it may chirp or beep enthusiastically, almost as if it is delighted to be with you, rather than simply rolling behind in silence.

Astro may lift its "head" and make a sound that suggests it is wondering, "What's next?" if you stop moving. Astro feels less like a tool and more like a participant in your home activities thanks to this form of interaction, which is further demonstrated by its entertainment capabilities.

Interactive Learning and Games

Astro has features that allow individuals of many ages to play, so it's not only about functioning. The capacity of Astro to involve people in games is one of its most notable entertainment features. Astro can take an active role in activities, whether you're just wanting to kill time or spend quality time with your family. Astro offers a variety of games for single players as well as multiplayer options. It may host interactive challenges, quiz nights, and trivia contests, for instance. Astro can draw from a vast collection of facts and inquiries to keep things interesting and novel because of its interface with Alexa. When you say, "Astro, let's play trivia," he will take up position and be prepared to ask you questions to see how much you know. It can even maintain score for several players, which adds organization and interest to gaming nights.

Astro can also play games like memory and pattern recognition, which are especially well-liked by children. The robot makes even the most basic games more immersive with its interactive display, noises, and motions. Imagine yourself in a Simon Says-style game where Astro gives you instructions and, when you follow through on them, rewards you with happy motions and beeps. When you don't follow her instructions, she playfully expresses disappointment. Astro is an excellent choice for families that include kids since it can help them learn via play in an engaging approach that blends cognitive and physical involvement. Astro can support these activities. It allows kids to learn and grow in their problem-solving abilities while still having fun. Additionally, kids find it simple to relate to Astro because of his upbeat personality and view the robot as a buddy rather than as a tool.

Dance and Music: Get Grooving with Astro

Astro can play music and even dance to it, which is one of its most endearing features. Astro may be your musical partner whether you're having a party, performing chores, or just lounging. Astro may access any music streaming service that is compatible with Alexa, such as Apple Music, Spotify, and Amazon Music, by using its connection with the voice assistant. Saying "Astro, play my newest playlist" will cause the robot to begin streaming the music of your choice right away.

Astro can dance, too, which adds even more enjoyment to this function. Astro can mimic dancing by tilting its head and moving its wheels in sync with the beat of the music. The robot may include a fun, interactive aspect into your music experience by swaying or rotating in tune with basic dancing.

Children will particularly like this function since they may dance with Astro and imitate its movements. Astro's dancing adds an additional layer of pleasure to music playback, whether you're having a dance-off in your living room or you just want someone to spend some time with while you listen to your favorite tunes. Although it's a tiny gesture, it helps Astro seem less like a passive speaker and more like a fun collaborator.

Additionally, Astro is able to identify specific commands pertaining to entertainment and music.

It moves along, dancing or swaying to the background music, and you can control it by asking it to pause, skip songs, or change the level.

Keeping Pets Entertained

Astro isn't just for people; your dogs may also take part in the excitement. Astro provides a great deal of enjoyment for pets, especially dogs and cats, because to its interactive features and mobility. Since many pets are inherently curious, Astro's mobility about the house is sure to pique their interest. While some pets would view Astro as a worthy companion, others might find its motions fascinating and follow it about. Astro can play light chase with your pet around the home or lead them from room to room. Dogs may follow Astro in this entertaining game, which can keep them occupied and active while you attend to other matters. While you're not home, Astro can completely assist keep pets engaged. You may remotely operate Astro by directing it to the various areas of your home where your dogs are located, all using the Alexa application. You may use Astro's speakers to communicate with them, keep an eye on them, and even play with them by moving Astro in ways that pique their curiosity.

Astro from Amazon is a good investment if you are a pet owner. Its sophisticated sensors let you to keep an eye on your pet's activities, including how much food and water they consume, and notify you via your smartphone if something is off. It can also provide treats remotely and play with your dogs. Here are a number of examples of how Astro offers careful pet care.
1. Behavior Monitoring: Your pet's behavior while you are away can be observed by the sophisticated sensors. Vital signs that provide information about your pet's health include food and water intake, activity levels, and sleep habits.

2. Remote Interaction: Astro's interactive capabilities let you interact with your pet from a distance. You can play with your pet by directing its motions and starting interactive activities with the companion app. It entertains and engages your pet while offering cerebral stimulation.
 3. Treat Dispensing: To encourage and reward good conduct, Astro may also administer rewards. By using the app, you can reward your pet's good behavior by activating the treat dispenser and giving them a special treat.

Personalized Entertainment Experiences
Astro's capacity to change with its customers' tastes and routines over time is one of its most captivating aspects. Astro picks up more information about your preferred forms of entertainment the more you engage with it. Based on your previous interactions, Astro can now recommend new games, activities, or entertainment alternatives to you. If you play trivia games with Astro on

a regular basis, for example, the robot may recommend a new question category that it believes you'll like. If you watch TV or listen to music frequently, Astro can recall your tastes and make tailored suggestions. This results in a more personalized entertainment experience, where Astro seems more like a buddy that knows your preferences than a multipurpose gadget.

Astro offers family members more customizing options. With the ability to store individual preferences for each member of the family, Astro can fluidly transition between various entertainments modes according on the person it is engaging with.

Personalized touches like playing a child's favorite song, recalling an adult trivia game, or bringing up the latest episode of a podcast enhance Astro's entertainment value.

Smart Home Integration: Full Control Over Your Entertainment

Astro's smart home connectivity allows it to manage your house's entertainment systems in addition to its built-in entertainment functions. Astro can communicate with any Alexa-enabled entertainment device, including smart TVs, sound systems, and streaming devices, because it is powered by Alexa.

For instance, when you're preparing for a family movie night, you may ask Astro to turn down the lights or start a movie on your smart TV. Astro does not require a remote control to change channels, adjust the volume, or switch between streaming services. Because of this, Astro serves as a practical central location for organizing all of your home entertainment, letting you enjoy yourself uninterrupted.

Because Astro is compatible with smart home systems, you could also use it as a controller for the atmosphere, lighting, and music at social events and parties. Astro can create the perfect atmosphere by adjusting the lighting and music depending on your choices, as well as providing entertainment alternatives like as games or quizzes to keep attendees entertained. Astro as a Social Companion

Finally, Astro's ability to be a social buddy is among the most intriguing components of its entertainment offerings. Astro is meant to be a member of a home, engaging with several people and enhancing shared entertainment experiences. It is not only for solitary usage.

For example, Astro may be a delightful host at get-togethers with the family or parties. It may keep the mood lively and promote interaction between visitors by getting everyone involved in group activities like karaoke, quiz games, or party games. Astro becomes a lively element of the social environment due to its flexibility and capacity to react to various individuals.

Astro is more than simply a smart gadget because of its capacity to follow people about and strike up a conversation, play music, or launch a game, even in more relaxed situations. It's a robot that can add joy and memories to times and makes you feel like a member of the family.

More than just a useful helper, Amazon Astro fosters connection, enjoyment, and fun in the house. With its joyful attitude and adaptable features, Astro is a unique companion for people of all ages, whether they are playing games, dancing to music, interacting with children, or even amusing pets. Beyond the capabilities of conventional smart gadgets, its interactive features provide a more dynamic and customized experience that enhances daily living.

For music enthusiasts: Enjoy your preferred songs at any time, or location within your house. Saying something like, "Astro, play music from the 1990s," or "Astro, play Aerosmith's "Don't Wanna Miss a Thing" on Amazon Music," is a good place to start.

For the news junkie: Astro uses simple voice commands to connect you to your preferred podcasts, whether they cover sports, world news, or captivating stories. To properly ensure that you don't miss a minute of your episode, say "Astro, play my latest podcast" and then "Astro, follow me".

For movie lovers: Astro offers access to all of your preferred streaming video services. Say "Astro, play me 'Lord of the Rings' on Prime Video" to catch up on all of your favorite episodes and films.

For the inquisitive mind: Let Astro assist you by suggesting that you read The Great Gatsby when you have the time to listen but not to read. You may listen to your favorite good books as intended by the author by syncing Astro with your Audible account.

You can Interact with Astro's entertaining personality.

Astro adds humorous moments to your everyday routine by combining sound and action. Try out these popular interactions, or come up with your own:

"Astro, what are you thinking about?"
"Astro, cross your eyes."
" Astro, do the robot."
" Astro, say something."
"Astro, give me a rap."
"Astro, what is your age?"

"Astro, which color is your favorite?"
" Astro, can you burp?"
"Astro, behave like a sheep."
" Astro, oink like a pig."
" Astro, act like a whale."
" Astro, how are you?"
" Astro, have a nice day."
" Astro, beatbox."
"Astro, how was your day?"
" Astro, play Robot Charades."

How Astro can entertain and educate kids through activities.

Getting the Children Involved: A Fun Companion
Astro was created with the full family in mind, which includes providing enjoyable and instructive entertainment for children. Astro's adaptive AI enables it to communicate with kids in more customized ways than only through games and music. Astro can recognize them by name, keep track of their favorite games and pastimes, and even recommend new experiences. As the kids start to think of Astro as their own personal robot buddy, this helps to strengthen the bond between Astro and the family. Astro could engage younger kids with educational activities. Astro can read aloud to you or participate in educational activities that focus on teaching letters, numbers, colors, and shapes. Because Astro is able to move, it may transform a routine story time into an engaging experience by acting out or moving specific scenes to make the story come to life. Imagine Astro

delivering the narrative in Alexa's voice while rolling about the room and acting like a character in the tale. Astro can assist older children with more complex learning, including arithmetic or spelling tests. Astro's quiz games can become increasingly difficult, keeping the kids interested as they increase their abilities. Astro can also provide tailored tests according on the child's areas of strength and growth because it remembers previous interactions, which makes learning more focused and efficient.

Because Astro is a lively being, it can also react to lighthearted inquiries and jokes. Astro will roll about excitedly as it delivers the punchline if you ask it to tell a joke. This characteristic gives the robot more personality and gives the impression that it's a humorous buddy who's always up to make things entertaining.

Astro is an avid gamer! Saying, "Astro, play hide and seek," will allow your children to engage in the timeless game of hide and seek. Remember to also check out the other entertaining games that are only available on Astro, such as Robot Charades and Dance Break.

The way that Astro, the house robot from Amazon, combines entertainment and technology makes it especially enticing to families with young children. Astro's clever features not only provide useful solutions for managing the house, but they also play a significant part in keeping kids entertained and educated with educational activities. Astro is a friend that is meant to be more than simply an aid; it may pique interest, promote education, and promote fun conversation. This chapter delves into how Astro may become a well-rounded part of the home by interacting with kids in meaningful, enjoyable, and instructive ways.

Astro's Child-Friendly Personality

Children can particularly relate to Astro because of its design. In contrast to all the conventional electronics, which can feel cold and unwelcoming, Astro exudes warmth and friendliness. It is more than simply a machine that obeys orders; it has a personality of its own. Children may relate to Astro's persona because of the way it tilts its head, makes noises, and utilizes its screen to show emotion. This greatly facilitates children's comfort level and encourages fun interaction with the robot.

Astro can enthusiastically answer inquiries from kids, recall their favorite games or hobbies, and welcome them by name. Astro's personality makes studying with the robot feel less like a chore and more like an experience, whether it's through games, storytelling, or interactive tasks.

Additionally, it fosters a great sense of friendship by making Astro feel like a reliable friend who is constantly willing to interact.

Logic Games and Math

Astro provides an assortment of informative and entertaining math and logic games. These games test children's critical thinking abilities, puzzle-solving abilities, and problem-solving techniques. Astro may assist younger kids build the fundamental math abilities they need by providing them with basic arithmetic quizzes or easy counting activities. When it involves challenging older kids' intellect, the robot can provide increasingly intricate arithmetic problems and logic riddles to win.

Because these games are interactive, kids are actively interacting with Astro, getting feedback, and gradually honing their abilities rather than not actively answering questions or doing puzzles.

Both storytelling and reading

Reading may be a challenging task for a lot of kids, especially those who are just beginning to learn. Astro can facilitate this process by adding enjoyment and interaction to reading. The robot as well has the ability to read aloud stories while using dynamic motions and sound effects to enhance the storytelling. Storytime is lot livelier and more fun when Astro is involved, whether it's a traditional bedtime tale or an instructive one.

Children can engage with the stories in a large variety of ways. Kids may interact more fully with the content when Astro stops the tale in the middle to pose comprehension questions. In addition, it can inspire children to play out certain scenes or make predictions about what will happen next, transforming reading into a participatory and dynamic experience. This fosters imagination and creativity in addition to helping with reading comprehension.

Astro can act as a quiz master for kids who are more proficient readers by posing questions concerning the tales they have read and encouraging critical thinking. This makes it a fantastic tool for improving vocabulary and understanding.

Science Exploration

Astro could also be used to teach science in an engaging and dynamic manner. Astro makes science interesting and approachable for children, from helping them through more difficult experiments to teaching basic ideas like the solar system or the water cycle. Children may ask Astro science-related questions, and Astro can include them in a more interactive learning process rather than merely providing a canned response.

For instance, if a young child inquires, "How do plants grow?" Astro is able to reply with an interactive lesson in addition to information. The youngster could be inspired to begin a little gardening project and explain each step along the way. Inquiring about the plant's growth on a regular basis allows Astro to reinforce lessons learned about photosynthesis, water cycles, and plant life cycles.

Children are better able to comprehend abstract ideas and maintain their curiosity about how the world functions because to these real-world connections. With its display that illustrates planets, stars, and galaxies while elucidating the fundamentals of astronomy, Astro can also assist children in their exploration of space. It might lead students on a virtual space trip where they "travel" to several planets and gain knowledge about space exploration, gravity, and distance.

Physical Activity and Movement

Astro promotes physical activity in addition to brainteasers and cerebral difficulties. At a time when devices frequently capture children's attention, this may be a big help for parents trying to keep their children active. Astro is indeed capable of guiding kids through basic workouts or dance routines. Astro's ability to move allows it to roam the room, inspiring kids to imitate its motions or dance to the music. For smaller children, who could view Astro as a friend they can imitate or run about with, this is very entertaining. Additionally, the robot may lead kids in stretching exercises that will help them release some energy and develop good habits.

Astro can lead games like "Follow the Leader" or hide-and-seek, transforming physical play into a social activity for families with many kids. These activities become more engaging and dynamic as a good result of Astro's capacity to move and recognize things, allowing it to roam the house.

Creative Play: Music and Art
Children may participate in creative play with Astro in addition to conventional learning activities. Astro is a friend that can help young musicians and artists develop their creative side.

Drawing Games and Art Challenges
Astro isn't a painter, but it may lead kids through painting tasks or sketching activities. For example, Astro may give the child a sketching suggestion to help them come up with something based on a subject. "Let's draw a jungle scene—what animals do you think live there?" is one possible sentence. When the youngster is finished, Astro can inquire about the drawing and encourage them to discuss their creative decisions. This enables children to reflect on their work and helps them think critically about what they have created. Astro could additionally instruct children in fundamental sketching methods by outlining specific steps that they can copy on paper. Even if it may not visually illustrate, Astro's directions are sufficiently

explicit to lead kids through basic sketches and promote the development of their drawing skills.

Music Exploration

Astro can support kids' musical exploration in a number of ways. Astro can explain fundamental musical ideas to younger children, such beats, rhythm, and melody. It can play various musical notes and ask them to determine which ones are high or low, or it could take them through clapping exercises to teach rhythm. Children who participate in interactive learning activities might improve their comprehension of music's structure and acquire an ear for it.

Astro can expose children who are a bit older to various musical styles or instruments. Astro can assist children in discovering their preferred genres of music by playing brief samples of techno, jazz, rock, or classical music. Astro can also explain the operation of various instruments and display photos on its screen to offer information about them.

Astro may also be a big help when it involves getting youngsters to join in on their favorite songs during karaoke sessions. After every performance, it may even provide supportive or entertaining feedback, which boosts children' self-esteem and gets them enthusiastic to sing. In addition to being enjoyable, this kind of musical involvement improves a child's voice and auditory development.

Personalized Playtime

The capacity of Astro to adjust to the tastes and learning style of each child is one of its most intriguing characteristics. Astro can adapt its activities to the child's interests since it remembers interactions and preferences. For instance, when a youngster plays mathematical games with Astro on a regular basis, the robot will start to recommend new problems in that domain. Astro may prioritize providing games with an artistic or musical theme if a child shows a preference for artistic endeavors.

Because of her individualized approach, Astro is a fun friend for kids of all ages and interests. Through identifying each child's interests, Astro can make sure that playtime is constantly engaging and novel. Because of its versatility, Astro may also grow with the kid, providing increasingly difficult tasks and activities as their abilities progress.

Astro's routines may also be customized by parents to include particular timetables or activities. For instance, Astro may be configured to play more laid-back games in the evenings or to participate in educational activities when it's time for homework. With this degree of personalization, parents can be sure that their kids are engaging in a healthy mix of enjoyable and instructive activities.

Collaboration and Social Interaction

Astro may also encourage youngsters to work together, transforming solo play into a communal activity. Siblings or friends can be encouraged to collaborate to solve problems or finish tasks by the robot by having them conduct team-based games or challenges.

This is especially good for social development since it instills in kids the value of communication and collaboration.

For example, Astro can start a scavenger hunt by providing hints that need children to cooperate in order to locate items across the house. Alternatively, Astro can divide children into teams for trivia games, fostering a fun, instructive, and competitive environment. This keeps several kids interested and fosters the development of critical social skills in a positive, encouraging setting. Parental controls and safety.

Of course, parents' first concern is safety when it in any way involves their children's use of technology. Astro comes with a number of parental settings that let parents keep an eye on and supervise their kids' robot activities. Screen time limitations, material filters, and a review of the games and activities Astro does with their kids are all customizable by parents. Astro can explore the house securely and prevent any potential mistakes or accidents thanks to its integrated camera and sensors. Astro will provide a good balance of learning and play, so parents can relax knowing that their children will be kept safe during playing.

Astro is a useful instrument in every family home since it can keep kids engaged with a mix of educational activities, creative play, and interactive games. It's more than simply a voice-activated robot; it's a customized, interactive playmate that can educate, amuse, and support kids in their development. Astro is a genuinely unique playmate for youngsters since it offers an experience that goes beyond what conventional toys or gadgets can deliver by combining fun with education.

The Function of Astro in Daily Life

The Amazon house robot, Astro, is intended to perfectly fit in with your daily routine. Astro, despite its seemingly simple appearance, is a very useful tool that may help with a variety of chores and improve the way you organize your daily activities. Beyond the features of conventional digital assistants, Astro provides useful solutions, such as schedule management, alarms, and reminders. This chapter explores how Astro can take charge of your everyday routine organization, ensuring that you stay focused without being overburdened. Astro's function as a personal assistant extends beyond housekeeping duties. It makes your everyday life better by offering you individualized support and keeping you organized. Let's examine Astro's function as a trustworthy personal assistant:

Using Astro to Set Reminders
Let's be honest: in today's hectic environment, it's easy to miss important details. Numerous little things need to be remembered during the day, such as taking your pills, writing a brief email, or picking up the dry cleaning. Astro acts as a dependable reminder system, which greatly simplifies the task of handling these little but crucial obligations.

Personalized Reminders
Astro has a great feature that lets you make customized reminders. It has the ability to remember things like "take out the trash" or "call Mom," in addition to recognizing your preferences and customizing reminders to fit your schedule. For example, you might have a regular daily pattern of going for a jog at 6 a.m., Astro can give you a gentle reminder to lace up your shoes and go out the door. It could totally remind you to consume a healthy snack or to drink a cup of water at the right moment if you really

want to spread out your daily meals and snacks. Astro could also help you stay afloat on your deadlines, appointments, and significant tasks. It can deliver proactive reminders to your smartphone or through audio notifications by getting to know your habits and preferences. As a result, you won't ever overlook a deadline or significant occasion. These reminders are quite simple to set. All you really have to do is ask Astro to remind you of the current task; the rest is handled by it. Additionally, Astro could be set up to remind you on particular days, at specified times, or even on a regular basis. This is especially useful for routine actions like finishing a weekly duty or taking your prescription.

Voice-Activated Setup
Astro's voice-activated feature makes it simple to schedule reminders. To establish a reminder, all you should do is provide a short voice command and no typing of text.
Saying something like, "Astro, remind me to check the oven in 30 minutes," can help you remember to do a crucial culinary task. With Astro's easy-to-use reminder system, you can effortlessly transition from one activity to the next since it will notify you when it's exactly needed. Astro's voice command function is quite flexible, allowing you to set up several reminders in a single day. Want to check up on a buddy later but also need to remember to pick up groceries after work? Astro can effortlessly manage both reminders, so you never miss a beat.

Alarms for Daily Tasks.

Astro's alarm system can assist you in keeping a consistent schedule, even though reminders are an excellent method to stay on top of certain activities. The varied and easily customizable alarm functions of Astro are useful for a large variety of tasks, such as tracking cooking times or getting up in the morning.

Wake-Up Call in the Morning

Astro may act as your very own personal alarm clock, gently yet effectively waking you up. Astro may be set to wake you up with a simple beep, your favorite music, or relaxing noises. Because of its mobility, Astro can even come to your bedside if necessary, guaranteeing that you always hear the alarm in the room. Astro can also be programmed to roll closer to your bedside or to gradually increase the volume of the alarm if you have trouble getting out of bed. This will give you that extra push to get up. Because Astro's sound is more like a friendly, gentle nudge than an obtrusive noise, some individuals find Astro's presence to be more reassuring than that of a traditional alarm clock.
Astro allows families to set up many alarms in various areas of the house. Astro can take care of it for you if you would need the kids up by 7 a.m. but don't want to trek from room to room.
In order to add a personal touch and make getting out of bed less of a nuisance, it may also provide polite reminders such as "Time to get ready for school!"

Managing Your Time for Tasks

Astro's alarm feature isn't only for waking up. You could also use it to stay on track during the day. Astro makes sure that you are remaining on schedule without feeling overwhelmed by deadlines, whether it's a reminder to start working on a project, a lunch break, or a transition from one activity to another.

Astro can handle both seamlessly if you're cooking, for instance, and require separate timers for different tasks, like baking a cake in the oven and boiling noodles on the stove. Say "Astro, set a timer for the cake for 45 minutes," and "Astro, set a timer for the pasta for 10 minutes," and Astro will keep track of both and notify you when they are finished.

Astro can help you manage your time by creating alarms for study sessions, breaks, and meeting reminders. This is especially useful for students and remote workers who find it beneficial to have their days planned. Astro will lead you through your schedule with timely alerts, so you won't have to worry about losing track of time or checking the clock all the time.

Organizing Your Calendar

The capacity of Astro to assist you in managing a hectic schedule is one of its best qualities. Whether you need to manage a family schedule, do several errands, or have back-to-back meetings, Astro can organize everything and provide you with information on a frequent basis.

Keeping Your Calendar Synced

Astro can be synced with a number of digital calendars, including Apple's iCal and Google Calendar, so it can keep track of all of your events and obligations. Astro can help you prepare for the day ahead, remind you of impending activities, and even reschedule chores when circumstances change after they are synchronized. Your digital and physical schedules will always be in sync thanks to this connectivity, so you won't have to worry about forgetting a crucial meeting or deadline. Astro can remind you to get ready for your meeting and communicate with you over when to depart for your appointments, such as a doctor's appointment at 3 p.m. and a meeting at 5 p.m. Astro may also be asked, "What's on my schedule today?" and it will provide you with a brief rundown of what happened during the day.

Organizing Events and Sending Out Invitations

Astro has the ability to intelligently manage your schedule by syncing with your calendar. For instance, depending on your interests and availability, you may recommend the best times for certain activities, create reminders, and book appointments. You can also use Astro's schedule management to assist you in organizing activities with other people. Astro may help with event setup, invites, and follow-up on RSVPs if you're organizing a virtual meeting or a family meal. To make sure everyone is aware of an impending event, you could even ask Astro to follow up with them. The simple integration of Astro with your calendar facilitates event preparation. Astro can handle everything from one location, saving you the trouble of managing several platforms and ensuring that you are ready for both business and personal obligations.

Establishing Priorities

It's critical to give certain jobs priority over others in a busy home or office. Astro can assist you in doing that by letting you prioritize particular activities or reminders. Astro will notify you more frequently of events that are designated as high priorities, so you won't miss anything crucial. Time management is another area where Astro can help. You may ask Astro to plan an uninterrupted hour of concentrate if you need it for personal or project-related tasks. Astro won't notify you of any additional unimportant duties during that period unless they are urgent. This degree of concentration reduces distractions and helps boost output.

Family Scheduling
A family with several individuals may find it difficult to manage schedules, but Astro may make things easier. Every family member may have a personal profile, and Astro will manage all of the obligations. Astro makes sure that everyone in the family is aware of what's occurring when, whether it's a family meal, a parent-teacher conference, or a soccer match for one child. Whether it's time to get ready for school, go to a weekend event, or get ready for a family movie night, Astro may send out reminders for family events. Without requiring frequent clarification, Astro's ability to adjust to each family member's schedule guarantees that everyone stays in sync.
Astro can also be used by parents to remind children to do their chores. When it's time for your child to take out the garbage or clean their room, Astro can gently remind them. These gentle reminders help kids keep on top of their tasks and teach them time management skills so parents don't have to step in all the time.

TaskDelegation

Astro's job delegation capability is yet another fantastic feature. Astro can be used to distribute chores to family members or roommates if you're running a busy home. When someone's turn comes to clean the dishes, for example, Astro could remind them of it. In order to guarantee that jobs are divided equally and without misunderstanding, Astro can also keep track of whose turn it is to finish a particular task. For families with older children or teens, Astro can also give a reminder about academics or extracurricular activities. Astro can manage those reminders in an impartial, non-intrusive manner, saving parents from having to follow up on a regular basis.

Astro can also help with task management in the office. Astro can remind team members of their obligations when you're in charge of a project and need to assign them, ensuring that everyone is held responsible for their work. This laissez-faire attitude towards handling obligations guarantees that nothing eludes oversight.

Astro as a Coach for Time Management

Astro does more than just program reminders and alarms. It may also act as a time management coach, assisting you in dividing complicated jobs into smaller, more doable assignments. Astro can assist you in creating a timetable for a large project, with minor targets set along the way to help you stay on track. Long-term tasks, such as homework assignments, hobbies, or even planning a family trip, benefit greatly from this feature. Astro can provide you very gentle nudges, encouraging you to maintain your focus while providing support and prompts. Being able to break down a major goal into smaller, more manageable chores will help you move forward steadily and avoid feeling overburdened.

Astro is more than simply a digital assistant in your everyday life; it's a necessary tool for maintaining organization in both your personal and professional life. Its ability to organize your schedule, set alarms, and manage reminders gives even the busiest days structure and control. Astro improves your capacity to concentrate on what really counts, which makes your everyday life easier and more productive. It does this by blending in smoothly with your habits. You can easily ask questions, set timers, make to-do lists, and even place online orders with Astro by utilizing voice commands. Astro can comprehend requests in normal English thanks to its AI skills. Additionally, it fulfills requests and responds accurately.

Maintaining and Customizing Astro

Make routines more helpful using Astro and Intelligent Motion.

Routines are Astro shortcuts that save you time by grouping together a series of operations so you don't have to request them separately. You can automate how Astro interacts with compatible Echo and smart home devices to save time. You can create your own or use a Featured

Routine, such as:
"Astro, enable the Good Morning Routine"
Astro will wish you a nice morning, tell you something new, and then play your Flash Briefing.
"Astro, enable the Start my Day Routine"
Astro will provide you with news, traffic updates, and other information.
"Astro, enable the Good Night Routine"
Astro will wish you a good night and play sleeping noises.
To set up Routines, launch the Alexa app and click on More, then Routines. You can personalize by establishing the routines that you find most helpful, such as morning rituals, reading times, daily stock market updates, and much more.

Astro, Amazon's friendly home robot, is intended to ease and improve your daily life, but its appeal goes beyond its fundamental functions. At the heart of its functioning is the capacity to be modified with routines and downloaded abilities, making it a highly adaptable and helpful companion for a wide range of activities. Astro may be customized to meet your specific requirements and tastes, such as waking you up on time, shutting out the lights, or amusing the kids. This chapter looks into the numerous ways in which Astro's abilities and routines may be set up and altered, giving you an in-depth look at how to get the most out of the robot in your house.

Astro's sophisticated routines feature lets you automate many operations with a single trigger or command. You already have a decent idea of how Astro's routines operate if you're familiar with Alexa's. However, Astro differs from Alexa in that it brings a tangible presence to your smart home ecosystem through its ability to move and its sensory capabilities.

What is a routine?

A routine is essentially a set of operations that are started by a single command or event. As an instance, you can create a "Good Morning" routine in which Astro plays your preferred news podcast, raises the blinds (if connected to a smart home system), wakes you up, and even provides you with a morning weather and calendar update. Rather of giving each of these orders separately to finish, a routine lets you start them all at once with a single command, such as "Astro, start my day."

You may make routines as basic or sophisticated as you wish. They may be anything from brewing coffee (provided your coffee maker is intelligent enough) to checking in with the security camera in your front yard, or they could be as simple as turning on your lights or operating your media center. Astro may be made to fit into your lifestyle in a way that is both efficient and natural by customizing these routines.

Triggering Routines

A routine can be initiated in a number of ways. You may program them to turn on in response to a voice command, throughout a specified period of time, or in response to an event—like entering a room or pressing your smart doorbell, for example. The included sensors in Astro increase the versatility of regular triggering. For example, Astro can sense when you enter the living room in the morning and initiate a wake-up ritual that includes turning on the lights, brewing coffee, and playing background music.

Even when you leave the house, Astro can start routines for those who have smart locks or sensors on their doors. Think about yourself as you're rushing to work: In order to assure you that the house is secure, Astro may automatically turn off any unnecessary equipment, lower the thermostat, lock the doors, and give you a brief check-in. This degree of automation greatly simplifies household management and lessens everyday stress.

Multi-Step Routines
Astro's routines are excellent since they allow him to combine numerous actions into one. It is possible to design routines with several phases that build upon one other. For example, after helping you wake up in the morning, Astro may turn on your bedroom lights, play your daily briefing, and even check the traffic conditions before you leave the house. You can customize Astro's routines to be as detailed as you wish. You can create a "wind-down" ritual for the evening that includes telling you what's planned for the next day, locking the doors, turning down the lights, and turning on your favorite calming music. Because these stages are linked together, you may concentrate on other things while Astro handles the tedious work of manually finishing each activity.

Customizing Astro's Behavior: Meeting Your Specific Needs
Astro has several skills pre-installed, but it truly shines when you start customizing routines and downloadable skills to give the robot a more unique personality. With customization, Astro seems more like a useful companion created just for you and your home than a pre-programmed device.

Establishing Personalized Routines

You can start by creating customized routines that work with your daily schedule and preferences to fully tailor Astro's behavior. Assume that your family enjoys a Friday night movie at 7 p.m. on a regular basis. Astro can be programmed to automatically dim the lights, switch on your home entertainment system, and turn down the volume on any smart speakers you have in your home. Astro may even be set up to notify you throughout a movie in the event of a weather alarm or a crucial communication from a loved one. Additionally, you may establish routines for particular family members. Because Astro can distinguish between different voices, it can carry out customized tasks according on the voice issuing the instruction. If you have children, you can set up a homework schedule for them. For instance, Astro may remind them to finish their tasks by 4 p.m., turn off any distractions like the TV, and play some relaxing background music to aid with concentration. This type of personalized behavior guarantees that Astro turns into a helpful resource for every member of the family.

Routines Based on Seasons or Events
Astro doesn't have to stick to his daily routines. Routines can be altered for holidays or special occasions. When throwing a holiday party, you might arrange for a routine to begin as soon as your guests arrive. Astro could set up the holiday lights, play up some music, and greet guests warmly at the door. Alternatively, Astro might

automatically turn down the thermostat and draw the curtains when the inside temperature rises in the midst of the summer, if that's more your style.

Even in more grave circumstances, like home security, event-based routines are applicable. Astro may be set up to keep an eye on your house while you're gone and to notify you if it hears or sees any strange activity. Astro's camera can detect anything unusual, and if it does, it can communicate with other smart security systems to lock doors or sound an alert in real time.

Getting Used to Changing Schedules

Because life is never predictable, Astro's flexibility is advantageous. Astro is simply and rapidly updated to reflect any adjustments you make to your program. If you begin working from home, for example, you may modify your morning routine to incorporate more reminders for work-related chores or block out time for lunch and breaks. Astro may also be used to incorporate new fitness routines into your schedule by alerting you when it's time to stretch or begin working out.

Astro is made to adapt to your needs, so changing routines is a straightforward procedure. You may use voice commands or the app to make changes while you're on the road. Astro ensures that these changes are smooth and that it continues to be an indispensable component of your daily activities.

Downloadable Skills: Increasing the Range of Astro's Potential

Astro may be improved with downloaded talents in

addition to routines. To increase the robot's capability, you may add these skills, which are basically applications. Astro can expand and adjust to your needs with downloaded abilities, whether they are for productivity, entertainment, or housekeeping.

Where to Find Skills

The same location where Alexa skills are available is also where Astro's talents are accessible: the Amazon Skills Store. To locate talents that meet your requirements, you may browse through categories like productivity, entertainment, leisure, and home automation. Once you've discovered a talent you like, all you have to do is download and install it, and Astro will add it to its repertory.

For instance, you may download a fitness skill that provides personalized training schedules if you want Astro to have a deeper grasp of fitness. Alternatively, if you enjoy trivia games, you may add a talent that lets Astro organize a family-friendly quiz. In addition to adding functionality, these downloadable abilities let you customize Astro's actions to fit your hobbies and schedule.

Commonly Used Skills for Everyday Use

Home automation and productivity are two of the most sought-after Astro skills. For example, Astro could be made to work with other smart devices in your house, such security cameras, lighting, and thermostats. Astro may be used as a central hub for managing every aspect of your smart home setup by creating routines that integrate these abilities.

Other trendy skills are around entertaining. Astro has the

ability to play games with the family, tell jokes, and even play music on command by downloading abilities. Astro is a more engaging presence in your house thanks to these enjoyable features, especially when you want to either relax or host visitors. Additionally, there are talents that support productivity. With Astro's downloadable productivity skills, you can keep on top of your tasks by managing to-do lists, sending reminders, and organizing your shopping list. With the help of these abilities, you may automate repetitive chores like making food lists before your weekly shopping trip.

Skills for Children
Astro also provides a large selection of skills created especially for kids. With the use of these abilities, which center on safety, amusement, and education, Astro can grow to be a dependable friend for the family's younger members. Astro can captivate kids and support their learning with engaging stories and instructive activities.
Astro's behavior may be adjusted by parents to make sure it fits with their child's hobbies and age. Astro may be programmed to remind your child to accomplish certain tasks, including brushing their teeth, doing their homework, or getting ready for bed. Additionally, there are abilities made to give youngsters positive reinforcement, which not only helps them form good habits but also adds enjoyment to everyday activities.

Optimizing Astro's Techniques and Practices
Astro has a ton of built-in features and downloadable talents, but its real value is found in customizing these features to suit your needs. You can customize Astro's behavior to ensure that it operates precisely how you

would want it to, whether you're using it to automate your house or to send out updates and reminders throughout the day. Changing the timings and sequences Changing the order and timing of the activities is one of the simplest methods to improve Astro's routines. For instance, you may spread out the various activities by a few minutes if you think Astro's morning routine moves a bit too quickly for your taste. Rather than everything occurring at once, Astro can wake you up first, then wait a little while before brewing coffee and giving you an update on the news of the day.
Similarly, you may arrange routines to trigger at different times according on your schedule, if you'd rather Astro to do certain things only on particular days. For example, if you work from home on Wednesdays and Fridays, Astro may set up a unique "home office" schedule that includes playing background music, altering the lighting, and reminding you to take breaks.

Experiments with new skills
It's a good idea to occasionally study new abilities for Astro in order to keep things interesting. You may try out several skill kinds to determine which suits your lifestyle the best, since the Amazon Skills Store is always expanding. Trying out new features can help you get the most out of your robot, whether it's a new game for the kids, a language-learning tool, or an advanced home automation function. By incorporating these additional abilities into daily activities, Astro continues to be a dynamic and adaptable friend who develops with your needs. To get the most of Astro and make sure it improves your life in significant ways, you should always explore

and tweak its behavior. You can customize Astro's behavior to suit your requirements and tastes thanks to its talents and routines. Astro becomes more than simply a house robot when routines are customized and additional abilities are downloaded; it becomes an indispensable tool for organizing and enjoying everyday life. Astro fits your lifestyle, giving you entertainment and company in addition to automating boring chores to make your days a little bit easier and more pleasurable.

Taking Control of Your Data and Privacy

Keeping track of data privacy has grown more important in a world where everything is connected. Being a cutting-edge home assistant, Amazon Astro creates a lot of opportunities to improve everyday living, but these developments also raise important concerns about data collecting and privacy practices. This chapter examines in detail the data collection methods used by Amazon Astro, their subsequent handling, and the privacy settings that users may choose to protect their personal information.

Like many other smart gadgets, Astro completely needs data to operate effectively. It can traverse your home, understand your tastes, and obey your every direction with the aid of this data. To balance convenience with privacy, however, you must be aware of what Astro gathers, how it utilizes this data, and what controls you have over this process.

The Fundamentals of Data Collection: Why Does Astro Collect Data?

Fundamentally, Astro gathers data in order to efficiently carry out its primary tasks. These features are meant to enhance the user experience by streamlining and personalizing your daily interactions. Let's examine the main justifications for why Astro must get particular data:

1. Navigation and Mobility: In order for Astro to travel around your house securely, it must get familiar with its layout. In order to do this, it maps out your location using integrated sensors and cameras to collect data about the physical surroundings. This enables Astro to locate you when you call for it, discover charging outlets, and avoid obstructions. For the mapping procedure to work properly, some temporary storing of your home's floorplan information is required.

2. User Interaction: Astro operates mostly through voice instructions. Astro analyzes speech input and listens for wake words, like "Astro" or "Alexa," in order to comprehend and carry out your requests. Many smart gadgets share this capability, which lets the assistant do things like play music, send notifications, and create reminders.

3. Learning Preferences: Astro collects information about your daily habits, preferences, and commonly used

commands over time. This enables it to tailor your experience even more. For example, Astro may figure out that you like a particular type of lighting or music at a particular time of day, and adapt without your input.

4. Smart Home Integration: Astro gathers information to work seamlessly with these systems if you use it to operate smart home appliances like thermostats, locks, or lights. Astro needs this in order to operate more efficiently throughout the house and interact with your other gadgets.

5. Monitoring Features: One of Astro's other features is the ability to watch over your house when you're not there. For you to use its surveillance functions, cameras and motion sensors must be used. These features may include remotely monitoring your house or informing you if there is any movement while you are away.

Which Data Does Astro Collect?

Gaining clarity on the precise data that Astro collects can help consumers feel more in control of the process. The many categories of data are broken down as follows:

- Visual Data: One of Astro's main parts is its cameras. They enable home monitoring services and navigation. Images or video footage taken while on patrol or doing activities might be considered visual data. However, until certain capabilities, like house monitoring, are activated, the visual data collecting is transient and mostly utilized in real-time.

- Audio Data: Astro is a voice-activated gadget that scans the environment for wake words and interprets audio information to carry out requests. Voice recordings can be reviewed and deleted by Amazon users upon request, while the device will normally only save voice data for brief periods of time.
- Location Data: Astro maps your house digitally so it can get around easily. Astro uses this location information, which is kept on the device, to determine the locations of rooms and objects.
- User Preferences and Routines: Astro collects data concerning your interactions with it, including your favorite wake-up times, music preferences, and daily reminders, as part of its smart assistant features. This enables the gadget to better meet your demands.

Privacy Issues: What Really Happens to Your Data?

Knowing what data is collected is only one piece of the jigsaw for many consumers. Where does this data go and who has access to it is the second crucial question.
To control the data that Astro collects, Amazon has put in place a number of security measures. Here's a deeper look at what transpires following the collection of data:

- Local Storage: Certain information is kept locally on the Astro device, such as your home's map. This means that information about your home's layout is not transferred to the cloud or shared with Amazon unless you expressly permit it for purposes like troubleshooting.
- Cloud processing: Information such as voice commands, preferences, and some other kinds of data can be handled remotely. This facilitates the assistant's comprehension of instructions and helps it perform better over time. On the other hand, Amazon offers transparency by letting consumers access and control speech recordings via the Alexa app.
- Encryption: All information sent to the cloud servers of Amazon is encrypted. Voice data and any communication with other smart device are included in this. Encryption makes sure that even in the unlikely event that data were intercepted, unauthorized access would be impossible.
- Restricted Access: Data kept on Amazon's servers is only accessible by authorized personnel. According to the firm, it follows stringent privacy rules and procedures and conducts frequent audits to guarantee data safety.
- Policies for Data Retention: Users can really decide how long their data is kept on Amazon. Through the Alexa app, you have an option to manually erase data at any time or to have voice recordings automatically deleted after three or eighteen months, depending on your settings.

Taking Charge: Personalized Privacy Settings

Making use of the adjustable privacy options that Amazon offers for Astro is one of the greatest methods to guarantee the security of your data. Here are some tips for handling various privacy-related issues:

1. Handling Voice Information
You can always go back and listen to or remove your voice recordings from Amazon. Either Astro's settings or the Alexa app could be used to do this. To guarantee that Astro performs voice instructions without saving any data, you can initiate the "Don't Save" option if you would rather not to keep any voice recordings at all.

To control your voice data:
- Open up the Alexa app, then select "Settings."
- Make sure you click on "Review Voice History" under "Alexa Privacy."
- You can configure auto-deletion options or remove specific recordings from this section.

2. Controlling the Camera Access
Although Astro's camera might be an effective home surveillance tool, you might not want it to be in continuous operation. Fortunately, there are options for managing the timing and operation of Astro's camera:

- Turn on Privacy Mode: By turning on Privacy Mode, you can always turn off the camera. With Astro's cameras and microphones turned off in this mode,

you can be sure that no data is being captured or sent.

- Control Home Monitoring: You may choose when the camera is activated if you use Astro for home monitoring. You may, for instance, configure the camera to only turn on when you're not at home. Furthermore, the Alexa app allows you to view a live stream from the camera and operate it in real time.

3. Tailoring Location Data
Astro uses location information to efficiently navigate your house. Although this is necessary for it to work, you are in charge of how this information is used and kept.

Local Storage: Astro's home map is kept locally on the device by default. If you decide you no longer want this data saved, you may also opt to completely clear this data or reset the map.

Custom borders: You can limit Astro's freedom of movement in your house by establishing no-go areas and custom borders. This is especially helpful if you would like to keep it out of rooms or private spaces where you don't want any data to be captured.

3. How to Apply Wake Word Sensitivity
   Astro uses wake phrases like "Alexa" or "Astro" to trigger its voice-activated functions. You have the option to change the wake word sensitivity if privacy is an issue. This lessens the possibility of accidental activations when you're conversing often or don't want to interact with Astro.

   To change the sensitivity of a wake word:

   Open up the Alexa app, then select "Settings."
Choose "Wake Word Sensitivity" under "Device Settings."
You can adjust how quickly Astro reacts to voice instructions here.

Features of Privacy for Families
In homes with kids or extended family, protecting personal information becomes even more crucial. Astro has a number of privacy features that are intended to

meet the needs of various users while upholding a high standard of security:

Parental Controls: To restrict the kind of content or interactions Astro can participate in, families with youngsters can activate parental controls. Age-based customization of these settings guarantees that younger users are shielded from objectionable information and unlawful behaviors.

Voice Profiles: Astro is able to adapt interactions according to the speaker by identifying various voice profiles. This implies that information is linked to specific users instead of being compiled. You may control each profile individually if you wish to restrict the information that Astro gathers for particular people.

Frequent upgrades and security fixes

Amazon is always trying to make Astro's privacy features better. Frequent software upgrades guarantee that any such vulnerabilities are quickly fixed. These upgrades come with additional privacy options, security patches, and improvements that provide consumers even greater control over their data. Enabling automatic updates for Astro's software is a smart approach in order to maintain its current. By doing this, you can be confident that your gadget is always running with the most recent privacy safeguards applied.

Providing Customization to Empower Users

Astro has several capabilities that may significantly improve day-to-day living, but with great power comes tremendous responsibility, particularly with regard to protecting personal information. Users may take advantage of the convenience of a smart assistant without sacrificing their privacy if they are aware of what data is gathered, how it's utilized, and what controls are in place for that process.

Customization is the key to keeping your data under control. You may customize Astro to fit your unique requirements and comfort levels with its many privacy settings and capabilities. Astro provides you the control over your personal data, including the ability to manage voice recordings, set camera limits, and change location data choices.

Data Sharing and Integrations with Third Parties
It's important to note how data is transferred across different services and what safeguards are in place if Astro interacts with third-party programs or devices. For example, if Astro is integrated with other smart home devices, such smart locks or security cameras made by other firms, Amazon would need to make it clear to customers whether or not their data is shared with other businesses, as well as how they may restrict or limit that sharing.

How Astro Manages Biometric Data (If Any)
Astro can collect and retain biometric data to customize experiences, depending on its capabilities (e.g., voice recognition, facial recognition, etc.). It is very important for users to understand how sensitive data is managed,

whether it is transferred to the cloud or retained locally, and how to edit or remove it.

Transparency Reports and Incident Management

A little section on how Amazon responds to security events and data breaches could be useful. For example, does Amazon publish reports on its openness about demands for user data made by governments or law enforcement, and how it responds to these requests?

Astro's Methods for Minimizing Data

Users who are concerned about their privacy may find comfort in knowing that Amazon only collects the data required for Astro to operate as part of its data minimization strategy. Users would feel more secure knowing that their data isn't being gathered excessively or needlessly as a result.

User Data Portability

If users decide to stop using Astro, they could also be curious about how simple it is for them to view, download, or move their data. Providing additional information about Amazon's data portability choices might increase control and openness over user data.

Troubleshooting and Common Issues

When using a complex technology like the Amazon Astro, troubleshooting and maintenance are unavoidable. While

Astro is intended to integrate seamlessly into your home, like with any technology product, occasional difficulties may develop. Fortunately, with a little know-how, most frequent problems can be quickly fixed, allowing you to keep your Astro performing at its peak.

Connectivity Issues: Keeping Astro Online.

Astro relies significantly on a consistent internet connection to perform properly, whether for voice commands, streaming, or other cloud-based functions. Should Astro abruptly go offline or stop responding, there could be an issue with the network. To troubleshoot:
Verify your home Wi-Fi connection: Check that your router is running properly and that your internet service is active. Rebooting your router might help to address occasional connectivity difficulties. Also, try to see if other gadgets in your home are having connection issues.
 Move Astro closer to your router: If Astro is too far away from your router, it may struggle to maintain a good signal due to walls or barriers. Moving Astro closer to the router or using a Wi-Fi extender to boost signal strength may typically resolve issues.
Restart Astro: If the issue persists, consider restarting Astro. You can accomplish this by holding down the power button for a few seconds and selecting "Restart" from the menu. This simple reboot typically resolves minor software faults or temporary connection issues.
Software Update: Ensure Astro's firmware is up to date. Outdated software might cause connection troubles. The device will normally alert you for updates, but it's a good idea to check the settings menu from time to time to make sure you're using the most recent version.

Battery & Charging Issues: Keeping Astro Powered

Astro is a mobile gadget that roams throughout your home; thus, battery life is important to its operation. If you discover that Astro isn't charging correctly or that its battery is draining quicker than normal, here are some troubleshooting steps:

1. Place the Charging port:

Astro automatically returns to its port for charging. If Astro isn't docking properly, the charging ports may be misaligned. Make that the dock is on a flat, solid surface that is clear of dust and debris. Make sure nothing is impeding Astro's path to the pier.

2. Inspect the Charging Contacts:

Both Astro and its dock contain metal connectors that connect to allow charging. Over time, dirt or grime may collect on these contacts, interfering with the charging process. To guarantee a good connection, gently wipe the contacts with a dry cloth.

1. Calibrate Astro's Battery.

   If Astro's battery % appears inconsistent or incorrect, it may need to be calibrated. Allow Astro's battery to drain entirely before charging it fully and continuously. Doing this every few months can help keep the battery readings accurate.

2. Avoid Overuse when Charging:

   Using Astro intensively while it is charging might produce heat and reduce battery performance over time. To keep Astro's battery healthy, let it rest while it charges on the dock.

The Amazon Astro is unable to locate its charger.

If Astro is unable to locate its charging dock:
Ensure that Astro has a very clear path to its charging station.
Make sure the device is clean to avoid any interference with navigation. Make sure there is no debris or dust on the front panel. Verify that nothing is on the ground obstructing Astro's path. Ensure that there are no closed doors in the room where the charger is located so that Astro can get to it.

Verify that the charger is at the proper place and hasn't been moved from the map.
Make sure the charger is on a level surface with five feet of room in front of it and at least one foot of space on each side.
Make sure that any Ring Alarm Motion Detectors are at least 15 feet away from the charger.
- Take Astro and place it in front of the dock with the screen facing away if it still won't go to its recharge. Return Astro to the dock gently so that it can charge. Restart the Amazon Astro after it has been charged.
- Try running through Exploration once again if Astro is still having trouble connecting to its charger.

Amazon Astro Is Not Recognizing My Face.

To resolve facial recognition errors, try setting up Visual ID again.
If Astro doesn't recognize you, make sure:
You aren't too far or too near to Astro (5-10 feet is optimal).

Astro only sees one face.
Astro's microphones and cameras are turned on
If Astro is still having trouble recognizing your face, consider removing and resetting your visual ID profile.
Swipe down from the top part of Astro's screen and choose Settings.
Select Profiles, then select the profile name you're experiencing difficulties with.

To erase the Visual ID profile, tap the (-) symbol.
After erasing, touch the (+) symbol to establish a new Visual ID.

Amazon Astro Is Triggering My Ring Alarm.
Troubleshooting suggestions for Astro with the Ring motion sensor.
Change the location of your Astro charging dock and ensure that it is not within 15 feet of a Ring motion detector.
In the Ring app, change the Motion Detection Mode to Medium. Select Alarm > Motion Detector, then Motion Settings > Medium Detection.
If your Astro device continues to trigger Ring motion detectors after changing Motion Detection to Medium Detection, configure an Out of Bounds Zone. This keeps

Astro at least 15 feet from your Ring motion detector. Astro cannot access Out of Bounds Zones while on patrol.

To help avoid false alerts when Astro is near Ring Alarm Motion Detectors, make sure they are mounted at the proper height. Install 1st Generation Motion Detectors at 7 feet height. Install 2nd Generation Motion Detectors at 7 feet 6 inches high.

Using the Ring app, set each Motion Detector to Medium Detection.

Make sure Astro's charger is at least 15 feet from any Ring Alarm Motion Detectors.

If Astro still receives motion alerts after completing these steps, you may use your Astro app to create out-of-bounds zones for Astro, keeping him at least 15 feet away from your motion detectors. Alternatively, you may change Astro's patrol settings in the Astro application to exclude any perspectives from your patrol that are close to motion detectors.

Finally, if you still continue to get false motion alerts caused by Astro, you could switch your motion detector to Low Detection mode in the Ring App. Using Low Detection in a business setting may result in missed detections and increase the time necessary to detect motion and activate Ring Alarm. We strongly advise you to set your motion detectors to at least Medium Detection.

Ring Events Are Not Being Investigated by Amazon Astro

Astro is having difficulty investigating Ring sensor events.

If Astro is not examining Ring Events, try these troubleshooting steps:

Ensure that all rooms are individually named in the Astro app. Choose Settings > Maps > Edit Map. To add names to the rooms, choose each one.
Make sure your Ring Alarm sensors and motion detectors are added to the Smart Home groups in your Alexa app.
The Ring Alarm and motion detectors must match the rooms in which they are physically installed.
To manage Smart Home groups in the Alexa app, pick Devices, then Groups, and finally the room where the sensor is located.
Turn "on" Ring Event Detection and Investigations in your Astro app's Settings > Monitoring > Smart Alerts.

Amazon Astro is frozen or unresponsive.
Troubleshoot your Astro gadget to fix responsiveness difficulties.
First, try to restart Amazon Astro.
If restarting fails, Factory Reset your Amazon Astro.
Note: A factory reset returns Astro to its initial or original factory configuration.
If you still have troubles, contact Amazon Astro Customer Support (available daily between 5 a.m. and 10 p.m. Pacific Time) via the Astro app:
Launch the Amazon Astro app.
Select Settings.
Choose Help and Feedback, then scroll to the Contact Us area. Please ensure that your device and accessories are present.
Navigation and Movement Issues: Making Astro Move Smoothly

One of Astro's most notable characteristics is its capacity to navigate your house independently. However, there may be times when it becomes stuck, has difficulty negotiating specific locations, or travels erratically.

How to Troubleshoot Common Navigation Issues:

1. Clear obstacles: Astro maps your house using cameras and sensors to avoid potential impediments. However, little things such as toys, wires, and furniture legs may not always be identified, causing Astro to become trapped. Make sure the paths are clear, especially in regions where Astro regularly wanders.

2. Recalibrate Astro's Sensors: If Astro is having difficulty identifying obstacles or moving smoothly, its sensors may require calibration. There is an option in the device settings to tune Astro's movement and obstacle detection. This method just takes a few minutes and can considerably enhance navigation performance.

3. Check for any Software Glitch: software problems might create movement concerns. If Astro travels in unusual patterns or gets stuck regularly, reset the device. If it doesn't work, you may try checking for software updates or resetting the movement patterns.

4. Avoid Confined Spaces : Astro is meant for large, open areas, so if your home has short corridors or tight corners, it may struggle to maneuver. You may use the mapping tool to set virtual barriers that restrict Astro from approaching these dangerous places. Try establishing out-of-bounds zones around any such possible barriers in

your environment so Astro is aware not to enter them. Try deleting your map and using the exploration and tour space options once again if Astro is still having problems traveling.

Tip: To keep sawdust, dirt, sand, and other debris out of Astro's operational area, clean the front panel of the device on a regular basis.

Audio and Video Issues: Providing Clear Communication

Astro's microphones, cameras, and speakers are essential for video calling, streaming content, and accepting voice instructions. If you're having problems with sound or video quality, here's what you can do:

1. Check the camera lens:
Dust or smudges on Astro's camera lens might cause hazy or unclear video images. To guarantee a clear image, gently wipe the lens with a microfiber cloth.
2. Adjust Audio Settings.
 If Astro's speakers aren't delivering clear sound or the microphone isn't picking up your speech properly, you most likely would need to change the audio settings. In the settings menu, you may adjust the microphone sensitivity and speaker level to better fit your surroundings.
3. Reduce background noise:
Astro's microphones are highly sensitive, although they sometimes suffer in loud conditions.
If you're having problems getting Astro to comprehend your orders or your voice seems muffled during video

conversations, consider turning down the surrounding noise or transferring Astro to a quieter environment.
4. Restart for Persistent Audio or Video Issues: If audio or video issues persist, restart Astro. A reboot can resolve many temporary issues with sound and visual performance.

Voice Command Issues: Increasing Alexa's Responsiveness

Astro incorporates Alexa for voice requests; however, it may not always answer as intended. Here's how to troubleshoot if someone is having difficulty understanding you or is just not replying at all:
1. Check Alexa's Wake Word:
If Astro isn't listening to your voice instructions, check that you're using the right wake word. You may check or alter the wake word in the settings menu to something more intuitive or unique, especially if you have other Alexa devices in your house.

2. Reduce Competitive Noise:
Astro, like most voice-activated gadgets, may have problems hearing orders in loud surroundings. When attempting to engage with Astro, make sure to keep background noise to a minimum.

3. Recalibrate Voice Recognition:
If Astro repeatedly misunderstands your orders, recalibrating its speech recognition algorithms may assist. You may teach Astro to identify your speech more

correctly by going through a quick setup process in the device's settings.

4. Update your Alexa app:
Make sure the Alexa app is up to current. Outdated versions may cause compatibility difficulties with Astro's voice functions. Checking the app store for updates is a simple solution that frequently addresses command difficulties.

Maintenance Tips:

Astro, like any electrical gadget, requires occasional maintenance to guarantee that it continues to function properly over time. Here are quite a few suggestions to keep your Astro functioning like new:

1. Regularly clean the sensors and cameras:
Dust and debris can collect on Astro's sensors and cameras, affecting its navigation and video quality. To keep these components in good working order, gently wipe them with a dry cloth.

2. Regular Software Updates
Software upgrades frequently provide significant speed enhancements and bug fixes. To keep Astro working optimally, check for and apply updates on a regular basis.

3. Store Astro Properly When Not in Use:
If you would want to not use Astro for an extended length of time, make sure to keep it properly. To keep the battery

healthy, keep it in a dry, cool area and charge it completely before turning it off.

4. Check for Physical Damage:
Astro travels about your home, and while it's built to last, it's always a good idea to check it for indications of wear and strain, particularly on its wheels and body. Addressing small difficulties early might help to avoid more serious problems later.

Contacting Support: When DIY Is Not Enough

Sometimes troubleshooting on your own isn't enough to fix a problem. If you've followed the procedures indicated above and Astro still isn't operating properly, contact Amazon Support. Here's something to remember:

1. Document Problem:
Before contacting assistance, make a note of the problem you are having and any troubleshooting actions you have previously attempted. This will expedite the support process and guarantee that you receive the appropriate assistance as soon as possible.

2. Use Amazon's Support Options:
Amazon has several options to contact customer care, including live chat, phone support, and email. Depending on the complexities of your problem, alternative solutions may be more suited.

3. Warranty and repair:

If Astro needs repairs or replacement components, see if it is still under warranty. Amazon's customer support can advise you on warranty coverage and the next steps to have your gadget repaired or replaced.

Advanced Troubleshooting with the Astro App:

Remote diagnostics: Many consumers may not be well aware that the Astro app supports remote diagnostics. When Astro detects a problem, the app may deliver real-time status updates, allowing users to properly understand what is happening.

Log Monitoring: The Astro app also keeps track of Astro's recent activity and orders. If there is a reoccurring issue, users can examine these logs to determine trends or triggers for Astro's behavior, which can be helpful before contacting customer service.

Customizing Navigation for Specific Rooms:

Room-particular Boundaries: In addition to avoiding obstructions, the software allows you to set particular "no-go" zones for Astro in specific rooms. This can help Astro navigate more effectively by reducing unwanted contacts with crowded or complex environments.

Data Backup and Device Reset:

Data Backup for Setting: Astro's personalized settings, routines, and timetables must be backed up on a regular basis. If you would need to do a factory reset, this backup will let you restore the configuration without having to start from zero.

The Factory Reset Procedure: If everything else fails, a factory reset on Astro is your final recourse. A thorough, step-by-step explanation on how to do this is definitely beneficial to users. We addressed that below.

Proactive Maintenance Ideas:

Monitor Wheel and Motor Wear: Astro's wheels and motors may wear down over time, especially if it moves regularly. Users should inspect the wheels on a regular basis for smoothness of movement and for any unusual noises that might signal motor problems.

Keeping Astro in peak condition.

With proper maintenance and troubleshooting, your Amazon Astro may continue to be a dependable, useful helper in your home. Understanding how to deal with typical challenges, like as connection, mobility, and audio quality, can help you keep Astro an effective and interesting part of your everyday life.

How to Factory Reset Amazon Astro.

A factory reset will return Astro to its original factory settings and delete any account information.
Note: Before doing a factory reset, Restart Amazon Astro or try other troubleshooting information earlier explained. The factory reset of Amazon Astro will delete all of your account information, including maps.
To Factory Reset Astro:
Swipe down from the topmost part of Astro's screen and click on Settings.

Click on Device Options.
Click on Reset to Factory Defaults.
If the device is not responding, click and hold down the volume up and volume down buttons for 15 seconds.
Note: After factory resetting Astro, you will need to go through the initial setup steps again.

Keeping Astro Up to Date

Keeping Amazon Astro up to date is critical to ensuring its operation, security, and overall user experience. Astro, like any other smart device, is largely reliant on software upgrades to increase performance, add new features, and assure its security within your home. In this chapter, we will look at how important it is to maintain Astro's software up to date, how updates are provided, and recommended practices for ensuring that your Astro performs at top performance.

Understanding the Amazon Astro Software Updates

Amazon Astro's upgrades are more than simply bug fixes; they are critical for improving the device's capabilities. Software updates, whether they improve navigation, add new capabilities, or address security issues, are crucial to keeping Astro up to speed with the newest technology breakthroughs.

Why do software updates matter?

Bugs Fixed: Astro, like any other technology, may occasionally have difficulties or malfunctions. Updates assist to resolve errors, improve stability, and ensure that the robot works as intended.

New Features: Amazon often introduces new features and upgrades to enhance the user experience. By keeping Astro up to current, you guarantee that you get the most out of the latest improvements that make it more versatile and efficient.

Security patches: Cybersecurity is a big risk for connected devices. Regular software upgrades shield Astro from any vulnerabilities, protecting your personal information and privacy.

How Updates Are Delivered

Astro updates are often given over-the-air (OTA), which means they are downloaded and installed automatically with minimal user effort. Here's an explanation of how the updating process works:

Automatic Update: Astro is programmed to automatically update itself whenever a new software version becomes available. This keeps the robot up to date without requiring manual action from the operator. Automatic updates often occur during periods of inactivity, reducing interruption.

The Notification System: The Astro app will notify you when a new update is installed or if there is a problem with the updating process. In some circumstances, you most likely may need to authorize the update via the app.

Wi-Fi Connectivity: Because updates are distributed via the internet, Astro must be connected to a reliable Wi-Fi network. If your network is unreliable, the update may fail or be delayed.

Ensuring Astro is Always Up to Date

While Amazon has built Astro to handle upgrades effortlessly, there are quite a few things you can do to guarantee that your robot is always running the most recent software and working best.

Stable Wi-Fi connection

A robust and consistent Wi-Fi connection is required for Astro to receive updates. Without dependable internet connectivity, updates may fail or take longer to download, reducing Astro's performance. Here's how to maintain optimal connectivity:

Ensure connectivity in Key Areas: Because Astro travels about your house, ensure sure your Wi-Fi connectivity is robust in all rooms where it functions. If your home has Wi-Fi dead spots, try installing a Wi-Fi mesh system to boost coverage.

Verify Connectivity on a Steady Basis: Use the app to periodically verify Astro's connectivity to your network. If the signal is poor, troubleshooting your Wi-Fi or relocating your router may be helpful.

Monitoring battery levels.
Astro need sufficient battery power to download and install updates, particularly bigger fixes that take longer to install. To minimize interruptions, ensure that Astro is charged or docked when upgrades are expected to occur.
Keep Astro on the Charging Dock: Astro's charging port isn't only for keeping the battery charged; it's also where the gadget often launches updates. Make it a practice to return Astro to its dock whenever it is not in current use, so that it will always be ready to update when necessary.
Before updating, check the battery status: The Astro app will alert you of low battery levels, so charge the robot before commencing any big upgrades.

What Happens During an Update?
When Astro updates, numerous processes happen in the background to guarantee that the new software is downloaded, installed, and activated without interfering with your experience. Let's consider the normal steps of an update:

1. Update Download
The update package will be initially downloaded by Astro from Amazon's servers. The download time might range from a few minutes to several hours, based on the size of the update as well as your internet speed.

Monitoring Progress: While Astro normally does not require any interaction during the download, you may keep track of its progress using the app. If the download is stopped, the app will warn you, and the update will resume automatically once the connection has stabilized.

Pausing Downloads: If you wish to delay an update for whatever reason, most updates can be stopped in the app, allowing you control over when it happens.

Installation Phase
When the download is finished, Astro will start installing the update. During this time, Astro may momentarily stop regular functions.
Inactivity Throughout Installation: Astro will most likely suspend its regular duties to ensure the upgrade procedure runs well. During this period, the app may display a notice stating that Astro is downloading updates and will continue shortly.
The Rebooting Process: After installation, Astro may reboot to implement the modifications. This guarantees that the update's components are correctly triggered.

3. Post-update checks

When the update is finished, Astro will conduct diagnostic tests to ensure that everything is operating properly. This procedure is generally automated and may include recalibrating certain sensors or reloading data.

Self-Diagnostics: Astro runs self-diagnostics after updates to ensure that everything is working properly. You might just see Astro halting for a time to examine its internal system.

Update Log in the App: Following the update, the app will provide a summary of the new features or modifications, allowing you to discover what's new.

What should you do if an Update Fails?

Despite the executed updating process, there may be times when an update does not go as expected. Knowing how to solve common update issues can help you save time and stress.

1. Update is not downloading.
If Astro is having trouble downloading an update, there might be various reasons:

Check Wi-Fi Connection: Make that Astro is still linked to your Wi-Fi network. You may use the app to check the current connection status and troubleshoot any difficulties.

Restart Astro: Sometimes merely restarting Astro will alleviate download problems. Turn it off and back on to refresh the system.

Restart the Router: If the problem persists, try rebooting your Wi-Fi router to restore a better connection.

Update is stuck during installation
If the installation process stalls or takes too long, you can perform the following steps:
Check the battery levels: Ensure Astro has enough battery juice to finish the installation. Dock it on the charging

station and wait for the battery to charge completely before attempting again.

Pause and restart Update: If an update appears to be stuck, pause the process in the app and then restart it after a few minutes. This frequently resets the installation procedure.

Factory Reset (Last Resort): As a last resort, you can conduct a factory reset to return Astro to its original settings and restart the update process. Make care to back up any unique settings first.

Optimizing Astro's Performance After Update

After Astro has been updated, you could take a few actions to guarantee that its performance stays optimal.

1. Re-calibrate Sensors

Some upgrades may change the way Astro navigates your house, necessitating sensor recalibration. This is especially significant in homes with complicated layouts or changing lighting conditions.

Sensor Check: To ensure that Astro's sensors are operating properly, you may use the app to perform a diagnostic check. This is especially helpful following navigation-related improvements.

Conduct a Test Lap: Following a significant upgrade, let Astro explore your house for a short while to ensure that it can still travel correctly. Keep an eye out for any unfamiliar actions or hesitations.

2. Examine New Features

The majority of upgrades will include improved functionality or new features. Investigate what's new and see how these features might enhance your Astro experience.

Skill Improvements: Updates frequently enhance current skills or introduce new ones. For instance, an upgrade may make Astro's facial recognition better or enable more seamless integration with smart home appliances.

Customized Routines: Stay alert for fresh approaches to personalized routines. Regular updates make routines more flexible and give you greater control over how Astro interacts with your home environment.

Update Astro's Accessory

Updates may occasionally also make Astro more compatible with external devices like charging stations, cameras, or sensors. Verify whether any connected accessories have firmware upgrades available since they might enhance functionality and guarantee compatibility with the most recent version of Astro software.

Staying Updated About Upcoming Changes

Since Astro is meant to be a dynamic tool, keeping up with upcoming upgrades is essential to getting the most out of it.

Amazon Announcements: For information on impending upgrades or new features coming to Astro, keep a watch on Amazon's official announcements.

Astro User Forums: You can join online forums or communities where other Astro users discuss their tips, techniques, and experiences with new upgrades. This can help diagnose problems early on or offer insight into possible new applications for Astro.

1. User-Requested Updates: Mention how Amazon prioritizes upgrades by often considering user input. Emphasize the value of participating in beta testing programs or giving Amazon direct input in order to influence next releases.
Overseeing Update Timetables: Even though updates are frequently automated, let consumers know how to choose when they happen, especially if they would rather schedule them during a downtime to minimize interruptions.
Privacy and Data in updates: Include a disclaimer stating that upgrades could involve modifications to privacy or data collecting practices. Urge users to periodically check for revisions to the privacy policy to make sure they are okay with the data being shared.
Role of Companion App in Updates: Stress the importance of the companion app (like the Astro app) in handling updates, resolving issues, and monitoring new features.
Updating Amazon Astro's software is not the only thing that has to be done. It's about maximizing its efficiency, guaranteeing security, and making the most out of a

perpetually changing system that's built to adapt and improve over time.

The Future of Astro

The field of home robots is developing quickly because of inventions like Amazon Astro. The idea that robots could help us out with everyday chores around the house is quickly becoming a reality, opening us a world of opportunities to improve our quality of life. Amazon Astro is one of the early prototypes of what house robots may offer, but the future of this technology is broad and bright. It's evident that household robots like Astro have a wide range of potential uses beyond what they can do now as we investigate how they can advance in the years to come.

Transforming Everyday Activities
The potential for future house robots to simplify our everyday tasks is one of the most intriguing things. Imagine a robot that develops into a genuine personal assistant—one that does more than just send out reminders or alerts. You can already set reminders and manage your calendar with the current version of Amazon Astro. Future versions, though, could include more sophisticated decision-making tools to manage intricate scheduling conflicts or dynamically reschedule your day in response to urgent changes or personal preferences. As artificial intelligence advances, it is possible that home robots may take proactive control of household chores like food preparation by making recipe recommendations

based on available supplies, dietary requirements, and time limits.

In the future, Astro may take care of a large variety of domestic tasks, including automatic laundry folding and grocery shopping support by placing online orders for supplies when they identify a need based on the consumption habits of the home.

In-House Medical Care and Wellness

The medical field is one of the most exciting areas for future robot usage. The aging of the global population will probably result in a major increase in the role that house robots play in helping the elderly and those with disabilities. Future iterations of the Astro might include improved monitoring functions, such as the ability to detect falls, track vital signs, and provide real-time emergency response, in addition to its current capacity to monitor the home environment and provide basic security and support services.

For instance, consider Astro working as a personal healthcare assistant. She would be able to remind patients to take their medications, monitor adherence to prescription regimens, and even notify family members or medical professionals if something appears off. Such capability may also include connection with smart home appliances, where Astro might communicate with health-related gadgets like blood pressure monitors, glucose monitors, and smart thermometers to measure a person's health data over time. Furthermore, as wearable technology develops, robots like Astro may work in tandem with wearables to monitor heart rates, sleep patterns, and daily activity levels. They might then provide feedback and recommend lifestyle changes to enhance overall wellbeing.

Social Connection and Companionship

With Astro's existing features, consumers can communicate with family members no matter where they are by using video calling and texting. But home robots of the future should improve these exchanges in more significant ways. Future robots may provide companionship in addition to connection, owing to their potential for increasingly sophisticated natural language processing and social intelligence.

Robots can communicate with people and lessen their sense of loneliness, especially for individuals who are housebound or live alone. They could do this by comprehending context, emotion, and human needs more deeply. Imagine a home robot that can sense when someone is depressed and can play their favorite music or provide consoling words to help them feel better. Through individualized encounters, these robots might offer companionship to elderly or immobile people, fostering a feeling of emotional connection.

Furthermore, Astro's future may enable more engaging family activities in terms of connection. Consider a situation in which Astro may act as the focal point of online family gatherings, incorporating augmented or holographic technology to bring out-of-town relatives into your living room for a more interactive and communal experience.

Improving Safety at Home

Although Astro already offers rudimentary security and home monitoring capabilities, there may be substantial improvements in this domain for house robots in the future. Soon, robots similar to Astro could have improved voice authentication, face recognition, and sophisticated danger detection capabilities. Imagine if a robot could recognize possible trespassers, warn you to unlawful entry, or even call the police in an emergency.

In addition, future robots may be equipped with advanced environmental sensors that can identify changes in temperature, air quality, movement, and even any fire or gas leak dangers. Your home robot may essentially transform into a mobile, intelligent security system that keeps an eye out for potential threats both inside and outside the house, protecting you and your family all day long.

Automation and Integration for Smart Homes

Astro's connectivity with Alexa and pre-existing smart home ecosystems is one of its biggest advantages. Future home robots may act as the hub for all home automation, effortlessly managing every element of the family, as smart homes become increasingly more networked.

Imagine a world in which Astro learns your preferences over time and modifies the lighting, temperature, and entertainment selections according to your habits and

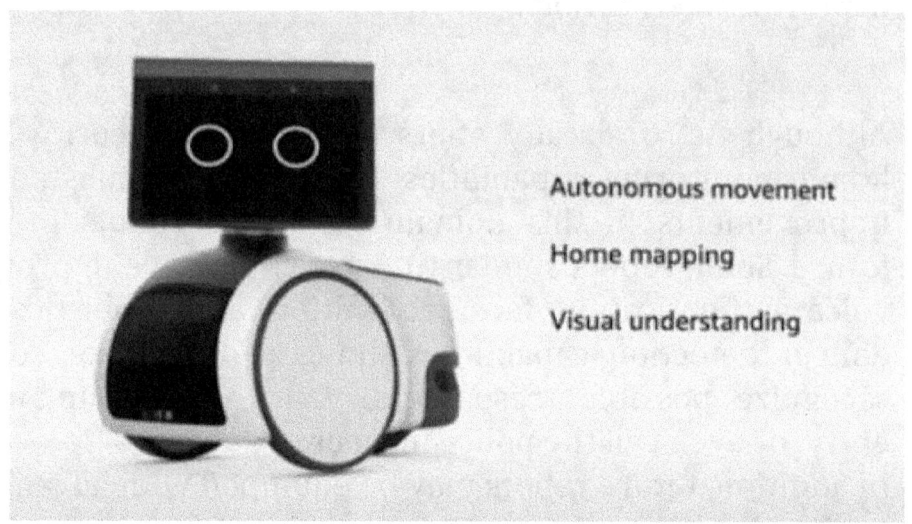

moods. It may also turn off the lights when you leave the room.

Robots like Astro may have the ability to manage increasingly complicated interactions with a variety of gadgets as houses grow more automated and intelligent. This may involve controlling energy use, making the most of temperature control systems in homes, or even liaising with self-driving vacuums or lawnmowers. Robots that can "talk" to one other and work together on projects might soon be a thing of the future, ushering in a completely automated home where many domestic activities can be finished without the assistance of a person.

Education and Personal Growth
Astro's ability to support education is a promising new direction for research and development. Through Alexa's

capabilities, Astro can currently provide basic interaction. However, in the future, robots may provide customized learning experiences depending on a user's learning preferences, age, or objectives. Astro might develop into a teaching tool for kids, helping with schoolwork, imparting new knowledge, and providing interactive learning through games or narrative.

The robot might act as an adult's personal coach to help them become more productive or acquire new skills. Future iterations of Astro might provide step-by-step guidance, specific recommendations, hints, and encouragement as you attempt new learning tasks like cooking, brushing up on arithmetic, or learning a new language.

Promotion of Emotional and Mental Well-Being

Home robots have the potential to be extremely important in promoting mental and emotional wellness in addition to physical healthcare. As artificial intelligence advances, robots like Astro could have a better comprehension of human emotions and be able to recognize and react properly to stress, worry, or melancholy. Potential therapeutic therapies provided by future robots include breathing exercises, guided meditations, and relaxing activities that assist people manage their mental health. These robots may also monitor changes in mental health, recognizing trends over time and making proactive recommendations for enhancing emotional wellness. Home robots have the potential to be useful instruments in supporting mental health, whether through mindfulness exercises or just by being a reassuring presence.

Personalization and Customization

The capacity to tailor robots to specific requirements is one of the most exciting upcoming breakthroughs in home robotics. Although Astro has many functions now, in the future users may be able to download unique behaviors, abilities, and even personalities for their robots. Would you want Astro to be more carefree and playful? Or would you rather have a more official, task-oriented helper? Users may be able to customize an experience with house robots in the future, according to their own tastes and way of life.

Moreover, Astro and related robots can pick up on interactions over time and get used to the routines, tastes, and personalities of your home. Imagine a robot that develops with your family, becoming more aware of each member's requirements and offering specialized care, all while assisting you.

Issues and Challenges with Ethics

It's critical to think about the ethical ramifications of home robot usage as we look to the future of these devices. Talks about privacy are already common, and as robots like Astro grow more commonplace, worries about data security, monitoring, and autonomy will also need to be addressed. In order to guarantee that the technology is utilized responsibly and does not violate people's privacy or personal liberties, future developments in home robots will need to carefully analyze these ethical issues.

Furthermore, the advancement of robotics might prompt concerns about the potential replacement of human labor in some areas, such security, cleaning, and caregiving. As the technology advances, it will be important to weigh the advantages of house robots against any potential social effects.

The Path Ahead

There is a bright future ahead for household robots like Astro. These robots might completely change every aspect of our lives, including healthcare, entertainment, security, and education. Home robots will likely become much more useful, individualized, and indispensable in our everyday lives as technology advances.

As one of the first of its type, Astro is a step toward a time when house robots will not just be a novelty but will play a major role in our daily lives and how we connect with the outside world. The potential of home robots is enormous, and we have only just begun to explore its possibilities.

Living with a Robot Companion

Living with a robot friend offers a novel way of integrating technology into daily life that was previously exclusive to science fiction. Robots like Astro from Amazon are becoming more than just useful appliances in homes. They integrate into our homes, engaging with our family, following our schedules, and even taking part in our everyday activities. However, incorporating this kind of technology into private, intimate settings raises difficult moral and societal issues that call for careful study.

Developing Connections with Robot Companions

The dynamics of our interactions with technology shift when we live with a robot. Living with a robot alters the nature of our interactions with technology. Unlike smartphones and voice assistants, which are mostly stationary and only respond to certain requests, a robot-like Astro travels around the house and interacts in more natural and dynamic ways. It feels more "alive," and for some, this might result in an emotional connection. While Astro is not created to replace human interactions, individuals may develop bonds to it as a consequence of its anthropomorphic mannerisms, including as facial emotions, noises, and movements that mirror real-life reactions.

For example, Astro may follow you from room to room, listening to your voice and even demonstrating a "personality" in its replies. For example, Astro may follow you from room to room, listening to your voice and even demonstrating a "personality" in its replies. This sensation of camaraderie may be soothing, particularly for the elderly or people who live alone. However, on a societal level, it raises concerns about the possibility of individuals replacing human relationship with artificial companionship. Could these robots eventually replace the need for genuine human interactions, and if so, how will this influence emotional well-being over time?

There's also the intriguing prospect that robots like Astro will improve social inclusion, particularly for those who struggle with face-to-face interactions, such as people with autism or social anxiety.

A robot companion can be a nonjudgmental presence that makes someone feel less alone while engaging in non-intrusive activities. However, this also implies that developers must be cautious of how robots are created, ensuring that they promote pleasant connections rather than isolating people from their surroundings.

Privacy and Security: Ethical Boundaries of Continuous Presence

One of the most pressing issues about the adoption of robot companions is privacy. Astro and other robots are outfitted with cameras, microphones, and sensors that allow them to explore and interact with their surroundings. While this is required for the robot to work, it also raises privacy concerns.

As the robot collects data, including video and audio, issues emerge regarding where it is stored, how it is utilized, and who has access to it.

The ethical ramifications are immense. Even though these robots are billed as home assistants, they are capable of capturing personal moments in everyday life. Whether it's an innocent family gathering or something more private, having a robot equipped with surveillance technologies begs the question of whether users have complete control over what is recorded and how data is distributed. In certain circumstances, the convenience of owning a robot may result in the loss of part of your privacy.

There is also worry about who owns this information. For example, may Amazon utilize Astro's data to better its services even if the customer hasn't specifically consented? Can third-party firms access this information? These issues underline the importance of robust, open privacy rules that safeguard consumers' rights and set explicit limits on data use. Furthermore, there is the problem of consent for guests to your house. Should you warn guests that they are being videotaped by a house robot, like you would with a home security camera? This raises ethical concerns regarding the level of control and openness necessary when others access your home environment.

Moral Issues in Decision-Making and Autonomy

As robots grow more intelligent and independent, ethical concerns about their decision-making capacities emerge. Robots like Astro are designed to be assistants, which implies they will take over some activities formerly handled by humans. However, as we give robots more control, questions emerge about where the line should be drawn.

Astro can make autonomous judgments on where to patrol in the house and which notifications to send? While these may appear to be minor difficulties, they reflect a bigger ethical challenge about how much autonomy we grant robots. If robots begin to make judgments that directly affect human lives, such as when to remind us of essential chores or who should have access to particular information, who is responsible if anything goes wrong?

Furthermore, when a robot's ability to operate independently in the house increases, it begins to question human authority in subtle ways. Even simple tasks like managing the thermostat or locking the doors may shift from a human's domain to a robot, and we will have to worry about our changing relationship with control and trust.

Accessibility and Social Equity in Robot Adoption.

One of the most promising societal implications of robots like Astro is their ability to democratize access to aid and care. Home robots may significantly increase freedom for the elderly, persons with impairments, and those with restricted mobility. Astro can transport tiny goods, make medicine reminders, and notify family members if anything strange happens. In this approach, home robots may minimize the need for full-time human caretakers or nursing facilities, allowing people to age in place in dignity.

However, access to this technology is not widespread. Robots such as Astro are still considered luxury commodities. This raise worries about the growing disparity between those who can afford to include such technology into their houses and those who cannot. If robots are to become a vital body of the family life, it is critical to investigate how they might be made more accessible to people from all socioeconomic levels, rather than being tools reserved for the wealthiest homes. As these technologies become more widely used, social justice considerations will become increasingly important.

Ethical Considerations for Job Displacement

As robots improve, there is a growing fear about job displacement in areas such as caregiving, household management, and even customer service. While Astro is unlikely to directly replace anyone's work today, the wider trend of household robots may have a direct impact on employment in industries where human labor has historically been required. For example, when robots get increasingly involved in supporting the elderly or providing home security, would there be a decrease in the demand for caretakers and security personnel? The ethical consequences of replacing human labor with robots are complicated. On the one hand, robots can save money while still providing continuous, 24/7 service.

The ethical consequences of replacing human labor with robots are complicated. On the one hand, robots can save money while still providing continuous, 24/7 service. On the other hand, there is a significant human cost when humans lose their employment or are replaced by computers. The goal will be to strike a complete balance between innovation and job preservation, so that workers in sensitive industries do not fall behind in the advent of robotic technology.

Emotional Attachment and Human-Robot Relationship

It's not only about practical ties between people and robots; emotional bonds are also developing when robots like Astro enter homes. While Astro is unable to perceive or reciprocate human emotions, its conduct is intended to elicit responses that simulate a sense of intimacy. Some individuals may find comfort in treating their house robot as if it were a pet or a companion. There are, however, ethical concerns about the emotional relationship that individuals may acquire to their robots. Is it good to create such ties, especially when the interaction is naturally one-sided? Will these bonds influence our interactions with actual people? Children, for example, may be more likely to build ties with robots due to their interactive and entertaining nature. As robots become increasingly incorporated into family life, we must carefully assess the consequences of emotional interactions with non-human partners.

There's also the opposite side: what happens if a robot fails or is no longer useful?
If humans have developed emotional links to their robot companions, they may feel loss or distress when the robot "dies" or fails. This is a relatively new domain for human emotions, and it poses an interesting challenge for engineers who must consider how robots "exit" our lives while minimizing emotional harm.

Long-Term Implications: The Ethics of Dependence

Finally, as we become more reliant on robots like Astro for daily chores such as managing home schedules and monitoring loved ones, there is an overall concern of how much we should rely on these machines. The benefits of having a robot remember appointments, track activity, and send reminders are apparent. However, there is a risk that we may become overly reliant on these technologies, therefore losing part of our capacity to govern our lives independently.

The dependence issue becomes more apparent for vulnerable people. For example, elderly people may grow so accustomed to robotic assistance that they lose their self-sufficiency abilities. Similarly, children who grow up around robots may develop the expectation that technology will always be there to help them solve difficulties. In both situations, there are ethical concerns regarding how much we allow robots to take over our everyday tasks and if this will have long-term consequences for human growth and autonomy.

1. Regulatory oversight: It could be worthwhile to investigate the role of government and regulatory authorities in ensuring that household robots, such as Astro, follow clear ethical principles. Regulations governing data privacy, consumer rights, and responsibility in the event of a malfunction or misuse will all have an impact on the future of home robot adoption.

2. Cultural Perspectives: People's comfort levels with robots differ across countries. For example, nations like Japan are often more accepting of robots in everyday life, but other civilizations may be more hostile owing to privacy concerns or societal preferences. Including a section on how cultural variables impact robot adoption might give readers a global perspective.

3. AI Bias and Fairness: Although Astro is not the most powerful AI system, more complicated household robots with decision-making skills may accidentally exhibit biases. It is critical to evaluate how these prejudices may

influence persons from various genders, races, and socioeconomic backgrounds. This is a bigger AI issue that is significant in the context of robots interacting with various homes.

4. The Evolution of Robot Capabilities: Discuss how rapid advances in AI and robotics may drive Astro and similar robots to more sophisticated functionality in the future, such as emotional intelligence or increased autonomy. This section might look into the possible advantages and hazards of such evolution.

5. Environmental impact: Robots are constructed with electrical components, many of which need rare materials and leave a carbon footprint during creation and disposal. Discussing how corporations like Amazon may reduce the environmental effect of robots may give an eco-friendly perspective.

Living with a robot companion like Amazon Astro presents both thrilling opportunities and serious ethical concerns. While technology promises to change the way we connect with our homes, care for loved ones, and live our lives, it also necessitates careful consideration of privacy, autonomy, social equality, and emotional ties. As we welcome robots into our homes, we must strike a balance between the convenience they provide and the ethical difficulties they provide, ensuring that they improve rather than degrade the quality of human existence.

How Astro Works with Other Smart Home Appliances

In today's fast-paced world, technology has effortlessly merged into almost every part of our lives, changing the way we connect with our surroundings. The notion of a smart home, in which networked gadgets collaborate to simplify activities and offer a personalized, easy living experience, has evolved from a distant fantasy to a present-day reality. Amazon Astro, a house robot, exemplifies this concept by not only providing its own distinct set of tasks, but also acting as a link between humans and their smart home ecosystems. Astro is not a stand-alone device; it is intended to improve and expand the capabilities of the gadgets currently in your home, working together to make daily living more intuitive and efficient.

This chapter investigates the different ways Astro interacts with smart home devices, offering insight into the technological basis that enables it. It also delves deeper into the potential future of linked life, in which robots like Astro might play key roles in controlling homes, workplaces, and even communities. Understanding Astro's role in this increasing ecosystem is critical to understanding its significance as more than just a curiosity, but as an essential component of the modern smart home.

Smooth integration with Alexa

Astro's close integration with Alexa, Amazon's speech assistant, is the foundation of its ability to communicate with other smart home appliances. Alexa acts as the digital hub of many homes, managing everything from speakers and gadgets to lighting and thermostats. As a mobile extension of Alexa's capability, Astro seamlessly integrates with the current framework by leveraging Alexa's capabilities.

Using the Alexa app or its built-in microphone, Astro can accept voice commands just like any other Alexa-enabled gadget. This implies that Astro can do whatever you can ask Alexa to do, including changing the lights, locking the doors, and regulating the temperature. However, Astro's mobility brings something extra: it can follow you around your house and deliver the smart home interface to any location you require. Imagine stepping into your living room and having Astro, without having to raise a finger, switch on the lights, pull down the blinds, and start playing your favorite music. Voice commands are now more convenient and follow you throughout your house.

Furthermore, Astro may serve as the main hub for increasingly intricate routines because it is designed with Alexa's whole feature set. For instance, you may program Astro to play white noise, switch on your security cameras, and lower the lights before bed.

Astro's mobility allows it to ensure that every room is securely locked or that, as part of a nightly ritual, all lights are turned out. With this degree of automation, a smart home's utility has advanced since automation now encompasses dynamic, individualized experiences rather than only static directives.

Security: On-the-Go Surveillance

One of Astro's most unique characteristics is its ability to function as a mobile security assistant. Even though many houses have sensors and static cameras installed, these devices are limited in what they can watch. However, Astro can patrol your house and provide you with real-time video feeds when it moves through different rooms. Astro adds m mobility and flexibility to an existing home security setup by integrating with third-party systems and smart security devices like Amazon Ring.

For example, Astro may react to guests by going to the front door and use its built-in camera to give you a much better look at who's there when it's paired with a Ring doorbell. Astro allows you to interact with guests just like you would with a doorbell. This capability comes in especially handy when you want to be able to monitor or greet guests even while you're not at home. An additional degree of security is provided by Astro's capacity to patrol at night or while you're away. It can inspect the condition of rooms or doors and provide live video feeds straight to your smartphone.

Because of its mobility, Astro can also help in an emergency. Astro may travel to areas of concern in the event of a fire or an intruder alarm, giving you and emergency personnel access to the most recent footage. Astro is a dynamic and adaptable addition to home security since it can change its location in response to changing circumstances, unlike fixed cameras that are constrained by their range of vision.

Managing the Weather: Fans, Thermostats, and Other Devices

Proper temperature control, which includes fan, air purifier, and thermostat control, is necessary for a pleasant house. Using Astro makes controlling the temperature in your house easier and more personalized. You can easily change the temperature or monitor energy use with Astro's seamless integration with smart thermostats like Nest or Amazon's own smart thermostat series.

However, Astro learns over time and does more than just follow your voice instructions or routines. Based on your daily routine and motions, Astro's machine learning algorithms may determine whether you prefer a lower or higher temperature. When Astro detects, for instance, that you typically withdraw to a colder room in the afternoon or when the sun strikes a certain side of the home, it may automatically reduce the temperature. Astro also offers control over fans and air purifiers that are linked to your smart home system, enabling a comprehensive approach to comfort and air quality management. Because Astro has sensors for both motion and air quality, it can turn on an air purifier to keep a space at the right temperature if it senses that a room is becoming stuffy.

Entertainment: Managing Your Home Media System

Astro is a mobile media assistant that elevates the entertainment experience. Astro can operate every aspect of your home entertainment system, including TVs, speakers, and even game consoles, thanks to its connection with Alexa and Amazon Fire TV. Let's say you're enjoying a movie night at home, but your remote is missing. No worries. All you would have to do is ask to Astro to play your preferred movie or launch your movie queue, and it will sync with your devices to accomplish this.

Multi-room audio enthusiasts can benefit from Astro's mobility in addition to its other advantages. Astro can go with you from one room to another room, playing music from your playlist via connected speakers or the device's internal audio. This eliminates the need for you to listen to music in just one area. Anywhere in the home, you may enjoy a seamless, continuous entertainment experience thanks to this function.

Additionally, Astro can manage the profiles and preferences of several users and serve as a center for family enjoyment. With individual Alexa profiles for each family member, Astro may customize the entertainment selections. When a youngster requests to watch television, for instance, Astro may make sure that only content suitable for their age is shown. Your smart home entertainment ecosystem gains even more convenience and control with this customized touch.

Energy Efficiency: Intelligent Lighting and Plugs

Another method to increase sustainability and convenience in your home is to integrate Astro into your energy management system. Smart outlets, lights, and plugs may all be connected to Astro to guarantee that energy is used effectively throughout your house. Astro can minimize energy consumption by turning off appliances and gadgets when not currently in use using routines or voice commands.

For instance, leaving electronics plugged in and using electricity when not in use is a widespread issue in many houses. Astro can keep an eye on these circumstances and either automatically turn off non-essential gadgets or notify you when action is needed. It may also be configured to lower the power output of specific devices during off-peak hours, which can help you save money on energy costs.

Another crucial component of energy control is lighting. You can easily manage smart lights from manufacturers like Philips Hue or LIFX with Astro's built-in Alexa connection. However, Astro does more than simply switch on and off lights. Astro uses routines and motion sensors to change the lighting in a room according to your presence. For instance, Astro may switch on your kitchen lights to a low level when you go in at night so that bright light doesn't interfere with your bedtime routine.

Astro's contribution to building an energy-efficient house is further enhanced by its ability to interface with energy storage and solar panel systems. It can keep track of the amount of energy produced and used, enabling you to run your house as sustainably as possible.

A Sneak Peek at the Future of Connected Living

much while Astro can do amazing things right now, as

house robots become more and more integrated into the fabric of connected life, the possibilities are much greater. The more gadgets and appliances that are made "smart," the more these items work together, and Astro will play a vital role in managing, coordinating, and communicating these products.

Healthcare is one sector that is ready for growth. Imagine Astro collaborating with medical equipment to keep an eye on family members' health and notifying caretakers or medical experts of any abnormalities it finds. Astro may be crucial to remote patient monitoring as the Internet of Medical Things (IoMT) develops, guaranteeing that health information and vital signs are sent in real-time to caregivers or medical professionals.

Advanced home automation is another exciting field where Astro might use artificial intelligence to use learned behavior patterns to predict requirements in addition to responding to orders. Astro's capacity to anticipate and react to environmental changes will revolutionize the way we conduct our everyday lives as smart homes become more user-friendly.

Furthermore, robots like Astro will have additional options as smart homes and cities become more connected. Home robots might interface with citywide networks to automate trash management and energy distribution as urban infrastructure grows more sophisticated. Astro might even help with home-to-car communication, vehicle preparation for when you depart, or synchronizing with public transit to facilitate more effective travel.

Creating an Ecosystem That Is Connected

Astro's contribution to the ecosystem of smart homes is only the start of how robots may improve connected living. With its capacity to carry out activities, communicate with other devices, and customize user experiences, house robots have the potential to surpass mere novelty and become essential components of our everyday existence.

1. Privacy within Smart Ecosystems: It would be helpful to include a brief discussion about Astro's approach to privacy in a networked smart home. Astro may give rise to privacy and data security problems if it interacts with other gadgets. The chapter might benefit from a discussion of Astro's role in preserving user privacy, data encryption, and openness about data sharing within the smart home ecosystem.

2. Third-Party Device Integration: Although Astro's compatibility with Alexa-enabled devices is well-established, it is worth noting that it also functions with third-party smart home devices, such as Google Home and Apple's HomeKit. This demonstrates how adaptable Astro is to different ecosystems that users may have previously established.

3. How AI Can Help Astro Increase Its Capabilities: Including a part on how advances in AI might enhance Astro's capacity for learning and making decisions. For instance, owing to advancements in AI, next upgrades may enable Astro to more precisely estimate your demands depending on your daily schedule.

4. Environmental Impact and Sustainability: Since sustainability is receiving more attention, it might be beneficial to talk about how smart home ecosystems, like Astro, can improve energy efficiency, lower carbon footprints, and encourage eco-friendly living.

5. User Customization: It would be more pertinent for individuals with particular preferences in their smart home ecosystem to include a discussion on the degree of control users have over tailoring Astro's routines and settings to meet their particular demands.

Devices like Astro will continue to develop and become increasingly important in our daily lives as we approach a day where technology permeates all aspect of our lives, including our work, pleasure, and daily lives. With Astro at its core, linked living has a bright future and is more dynamic, convenient, and intimate than it has ever been.

Astro Safety and Compliance Information

WARNING: In order to avoid hazardous situations for users and anyone in close proximity, responsible use is essential. Operate the device only in the environments that are listed in this handbook. Ensuring that the gadget is used within its specifications, intended usage, and appropriate surroundings is the responsibility of the user. To guarantee that the usage of this device does not endanger anybody, it is the user's obligation to make sure that onlookers are made aware of its presence and instructed in safe conduct around it.

Avoid listening at maximal volume for long periods.

An extended period of high volume use of the device may cause ear damage to the user. Users should avoid prolonged high volume listening in order to protect their hearing.

While Getting Ready to Use Your Device:

Even though the lasers are Class 1 (safe for the eyes), it is advised that you avoid staring directly at the laser light.

Use only indoors in dry areas. Only use your Amazon Astro ("Device") inside in dry areas to reduce the danger of electric shock. Keep your gadget out of steamy, very hot, or extremely cold conditions. For instance, keep your device and its accessories away from any heat sources, such as stoves, radiators, space heaters, heater vents, and other heat-producing objects.

Use your device and its accessories in an environment where the temperature stays between the range specified in this guidance for the device's operating temperature. When using your gadget often, it could grow heated. The device should not be utilized in a commercial setting; it is strictly for personal use.

WARNING: Keep tiny pets and children under three years old away from this device. Small children may be at risk of choking on tiny parts included in your device and its attachments. This gadget isn't a toy. Never stand, sit, or ride on this apparatus. Make sure that no one is put in danger when the item is operating before using it in a residence with people who have diminished physical, sensory, or mental capacities.

Make sure the setting in which you utilize Amazon Astro is suitable. Make sure your workspace is suitable for using the gadget in order to reduce the chance of accidents and property damage. Use caution while installing this product in homes that have sunken regions, descending stairs, or small steps that are 3.5–10 cm (1.375"–4") height with curved or cornered edges. Use of the gadget is not advised on shaky moving surfaces like moving walkways or elevators. Use the gadget only on ADA-compliant flooring. Use caution when using the gadget in potentially dangerous areas like oil spray, explosive gas, or delicate objects. In areas where there are combustible dust or particles, do not use the device.

When there are loose materials on the floor, such as clothes, loose documents, power cords, blind or curtains, or other delicate objects, do not let your device function in that area. If your device comes into touch with items while it is in operation, the objects may be knocked down. Make sure nothing delicate is positioned where it may be damaged in the event that the device's mast is lifted. Objects that come into contact with the mast might fall and cause injuries.

If an accessory or gadget is damaged, do not use it. If any of your device's attachments are cracked, broken, or otherwise damaged, do not use them. Take off any protective films prior to using the Amazon Astro. Before using, take off any protective films. If this isn't done, the accuracy of the sensor will decrease, perhaps leading to severe injuries.

Placing Objects on Your Device

Avoid covering your gadget or its attachments with blankets or other items. There's a chance of overheating if you do this.

You can put items in your cargo container that weigh up to 4.4 pounds. Nothing heavier than this should be put in your cargo bin. Avoid submerging your device's cargo bin in liquid, and don't put open cups or other things that are filled with liquid within the cargo bin to reduce the danger of damage. Make sure everything in the cargo bin is properly fastened to prevent unintentional collisions or property damage. Avoid storing anything that can be quickly expelled within the cargo bin of the gadget.

Avoid putting heavy things on other parts of your smartphone. It might shatter if you do that, endangering your safety.

Avoid covering up any sensors with cloth, stickers, or other materials. If this isn't done, the accuracy of the sensor will decrease, perhaps leading to severe injuries.

Avoid unexpected contact with device

When using and cleaning the device, make sure that fingers, body parts, hair, and items like strings, wires, loose clothes, and jewelry are kept away from it to reduce the chance of entanglement injuries.

When your gadget is moving, keep your fingers and body away from it to prevent unintentional touch. When the gadget is moving, do not lift it. Use the device's handle alone to raise it. Avoid sticking your fingers or other things into any other openings on the device or any of its attachments.

Keep an eye out for moving devices. Pay attention to your device's presence to prevent trips and falls. Make careful to alert individuals to the need to keep their heads and faces out of the periscope of the working equipment and away from any movements of the gadget. Do not run the gadget off a cliff, drop it, or use it carelessly in any other way.

Safety of Batteries

HANDLE WITH CARE. There is a lithium-ion battery that may be recharged in your Astro gadget. Reassemble, open, crush, bend, distort, pierce, shred, or try to get at the battery in any way. The battery should not be altered or remanufactured, nor should it be attempted to put foreign things into it. It should also not be submerged in liquids or exposed to fire, explosion, or other hazardous conditions. Utilize the battery exclusively with the designated system. Using an unapproved battery or charger puts you at risk for a fire, explosion, leak, or other problem. Avoid shorting out the battery or letting metal things that conduct electricity come into touch with the battery connections. Steer clear of dropping the battery or gadget. Don't try to fix the gadget or battery if you think it may be damaged after dropping it, especially on a hard surface. For support, get in touch with Amazon Customer Service. Especially whether in use or charging, keep the gadget and its charger away from heat sources and in a well-ventilated environment.

HOW TO USE AND CARE FOR YOUR DEVICE

Avoid exposing to fluids. Keep your gadget and its attachments away from liquids. Use caution when using your device or its accessories in moist areas. Be careful not to spill any drink or food on your accessories or device. Avoid placing any open item that contains liquid, such a cup or vase, on top of your gadget or in close proximity to it. When using your device again after it has gotten wet, make sure you switch it off gently and avoid getting your hands wet. Also, give it time to dry entirely. When using your accessories again, make sure they have thoroughly dried and disconnect all wires that may have become wet. Never try to dry your device or its components using a hair dryer or microwave, or any other external heat source.

Keep Your Device Clean
- Use clean hands only to handle your device and accessories. Before cleaning your gadget, secure any stray hair or clothes.

- During use, the device's and the charger's charging connections may get heated. Before cleaning, the operator must give the system time to cool down.

- Before cleaning, always switch off the device. Avoiding this might lead to a pinch danger.

- Unplug and unplug the power cord to avoid unintentional power-on during cleaning.

Use of Accessories

Only use the accessories that came with your device to supply power. Make sure there is enough room for your device where you are placing your charger. A downhill staircase should be at least three feet away from the charger. Make sure the charger and cord are not sticking out in places where people may trip over them. This will help prevent tripping hazards. Place your charger in a socket-outlet that is conveniently close to the devices that will be powered by or plugged into it. The charger should not be pushed into a power outlet. It is not recommended to plug several electronic gadgets into one wall outlet. When a socket is overloaded, it may overheat and catch fire.

Using third-party accessories might affect how well your device works. The limited warranty on your smartphone may be voided under certain conditions if you use third-party accessories. Before using any attachments with your device, make sure you have read all of the safety guidelines.

Servicing Your Device

Do not try to disassemble, alter, puncture, distort, or repair your device or its attachments in order to lower the possibility of harm to yourself or damage to your property.

Additional Safety Considerations

- Avoid submerging the cord or charger in liquids.
- Stop using the gadget, charger, or cable right away if anything seems broken.
- During a lightning storm, stay away from your gadget and any connected cables to reduce your chance of electric shock.
- Avoid subjecting the gadget to continuous magnetic fields. Strong magnetic fields have the potential to harm the gadget.
- Please refer to the user manual and any other material supplied by the manufacturer of any smart home device linked to your device for information on safety, compliance, and other matters.

Using Your Gadget Near Other Electronic Gadgets

When used improperly, your device may interfere with radio communications and electronic devices because it produces, utilizes, and emits radio frequency (RF) energy. Entertainment systems, personal medical devices, and electronic operating systems that are not properly placed or insulated from external radio frequency waves might all be impacted. If in doubt, contact the manufacturer as the majority of contemporary electronic equipment is insulated from external radio frequencies.

To find out if your personal medical devices—like hearing aids and pacemakers—are sufficiently protected from outside radiofrequency waves, speak with the manufacturer or your doctor.

FCC Compliance Statement

This device and its connected accessories, such as the charger (the "Products") are in accordance with FCC Rule Part 15. The following two requirements must be met for any product to function: Each product must:
1. Do not interfere with other products in a damaging way.
2. Accept any interference that is received, even if it might interfere with an intended function.

Note: In accordance with section 15 of the FCC Rules, the Product has been tested and determined to be in compliance with the restrictions for a Class B digital device or external switching power source. In a home installation, these limitations are intended to offer an acceptable level of protection from detrimental intervention. When not installed and used in compliance with the instructions, the product may cause detrimental interference to radio communications. The product creates, utilizes, and has the potential to emit radio frequency energy. But there's no assurance that interference won't happen in a certain setup. The user is advised to attempt correcting any adverse interference to radio or television reception that the product may cause by turning it on and off. This can be done by taking one or more of the following actions:
- Rotate or move the antenna that receives signals.
- Extend the distance between the receiver and the equipment.

- Attach the equipment to a separate circuit outlet than the one the receiver is attached to.
- Seek assistance from the dealer or a qualified radio/TV technician.

Details about Radio Frequency Energy Exposure

The radio technology employed in the Products has an output power that is less than the FCC's restrictions for radio frequency exposure. However, it is recommended that this device be used in a way that reduces the possibility of human contact while it is functioning normally.

When using this gadget, make sure there is at least 20 centimeters between your body and the radiator.

ABOUT THE AUTHOR

Markos Fletcher is a seasoned technology specialist with more than 20 years of expertise in the dynamic consumer technology field. Markos grew up in Seattle, Washington, and was exposed to the early wave of technological innovation in the Pacific Northwest. He developed a lifetime enthusiasm for knowing how things function at an early age, starting with disassembling home items and growing increasingly fascinated with technology.

Markos obtained his degree in Electrical Engineering from Stanford University, where his significant interest in robots and artificial intelligence took form. During his early career, he worked for some of Silicon Valley's most well-known tech businesses, contributing to innovative initiatives in smart devices, home automation, and AI-driven solutions. He has established himself as a progressive innovator over the years, and his name is on several patents related to smart home technology.

Besides his occupation, Markos is a well-known critic and observer of consumer technology goods. For those interested in learning about the newest developments in robotics and smart home technology, his blog, "The Connected Home," is a great resource. Markos, who is well-known for his frank and thorough assessments, gains the respect of both readers and other tech enthusiasts by approaching every product he reviews from a practical, real-world standpoint.

When Markos isn't buried in testing the newest technology, he likes to hang out with his family. His smart house serves as both a sanctuary and a testing ground for his most recent technological endeavors. He resides there with his wife, Sarah, and their two kids, Ethan and Lily.

Markos is a strong proponent of striking a balance between life and technology. When he's not using technology, he frequently finds happiness in cooking for his family, hiking the trails in the Pacific Northwest, and experiencing the great outdoors.

www.ingramcontent.com/pod-product-compliance
Lightning Source LLC
Chambersburg PA
CBHW071027240526
45469CB00006BD/2119